Inside NASA

New Series in NASA History

Before Lift-off: The Making of a Space Shuttle Crew
by Henry S. F. Cooper, Jr.

The Space Station Decision:
Incremental Politics and Technological Choice
by Howard E. McCurdy

Exploring the Sun: Solar Science since Galileo
by Karl Hufbauer

Inside NASA: High Technology and Organizational Change
in the U.S. Space Program
by Howard E. McCurdy

Howard E. McCurdy

Inside NASA

High Technology and Organizational Change
in the U.S. Space Program

The Johns Hopkins University Press

Baltimore and London

© 1993 The Johns Hopkins University Press
All rights reserved
Printed in the United States of America on acid-free paper

Johns Hopkins Paperbacks edition, 1994
9 8 7 6 5 4 3 2

The Johns Hopkins University Press
2715 North Charles Street
Baltimore, Maryland 21218-4363
The Johns Hopkins Press Ltd., London
www.press.jhu.edu

The quotation on p. vii is from Advisory Committee on the Future of
the U.S. Space Program (Norman R. Augustine, chairman), *Report of the
Advisory Committee* (Washington: Government Printing Office, 1990), 16.

Library of Congress Cataloging-in-Publication Data

McCurdy, Howard E.
  Inside NASA : high technology and organizational change in the U.S.
space program / Howard E. McCurdy.
      p.   cm. — (New series in NASA history)
  Includes bibliographical references and index.
  ISBN 0-8018-4452-5   ISBN 0-8018-4975-6 (pbk.)
  1. United States. National Aeronautics and Space Administration—
Management.   2. Organizational sociology.   3. Corporate culture.
4. United States. National Aeronautics and Space Administration—
History.   I. Title.   II. Series.
TL521.312.M33   1992
353.0087'78—dc20   92-18753

A catalog record for this book is available from the British Library.

This work relates to NASA Contract No. NASW-4236.
The U.S. Government has a paid-up license to exercise all rights
under the copyright claimed herein by or on behalf of the Government.
All others rights are reserved by the copyright owner.

To the memory of my mother,
Jo Janelene Test McCurdy

Space Is a very unforgiving place.

The Augustine Report, 1990

Contents

Preface

This book describes the transformation of the National Aeronautics and Space Administration, the institution created by the U.S. government in 1958 to carry out the nation's civilian space program. It relates how civil servants in that newly created agency appropriated traits from the National Advisory Committee for Aeronautics, the Army Ballistic Missile Agency, the Naval Research Laboratory, and the Air Force Ballistic Missile Program—groups out of which NASA was formed—and created an organizational culture supporting exceptionally high levels of performance. Employing that culture, NASA employees accomplished some of the most technologically difficult programs ever undertaken, including the first expeditions to the Moon.

Inside NASA further recounts what happened to the civilian space agency when, at the conclusion of the first full decade of space flight, the impetus for space projects waned, public and political support faded, and budgets for space exploration declined. It examines the rise of institutionalized bureaucracy within NASA as the space program matured and explores the implications of this experience for the management of high-technology programs in the public sector. In sum, the book traces the development of NASA from its creation in 1958 until 1992.

The book does not attempt to judge the merit or fault of NASA's transformation, except to observe how difficult it is to maintain an organizational culture like the one that NASA civil servants originally forged given the forces normally at work in government. Neither does the book dismiss the possibility of institutional revival. As the creation of NASA demonstrates, innovative cultures do arise. It does seem hard, however, to maintain such norms and practices for any appreciable period of time.

The book relies upon interviews, a survey questionnaire, and archival data to systematically identify NASA's underlying norms. Stories that people tell also reveal important facets of their behavior. Such stories are not always objectively true. Where possible, the book identifies the objective truth behind these stories, while respecting the fact that an event for which truth has been modified may yet contain a powerful lesson.

The research for this book was supported by the History Office of the National Aeronautics and Space Administration under contract NASW-4236, the purpose of which was to provide a history of NASA's organizational culture. While my description of NASA contains views widely held by persons associated with the American space program during the first three decades, the presentation of findings and conclusions found herein are strictly mine and do not necessarily reflect official NASA policy. As is the practice with all NASA-sponsored history projects, the author retained full editorial control over the contents of the book and thus is solely responsible for its accuracy and all judgments contained herein.

Acknowledgments

Many people assisted me in the preparation of this book, but no one more than Dr. Sylvia D. Fries (now Kraemer), who was Director of the NASA History Office when this study began and was promoted to Director of the Office of Special Studies in the Office of the NASA Administrator while the study was under way. She encouraged me to undertake this work and provided constant intellectual support as well as detailed critiques of my work as the study progressed. Dr. Roger D. Launius took over as NASA's Chief Historian in 1990 and continued the tradition of an intellectually challenging history program. This book went through many revisions under his guidance. Lee D. Saegesser, Archivist for the NASA History Office, guided me expertly to words and facts from the past, helping to satisfy my requests for further detail.

Among the employees of the National Aeronautics and Space Administration, present and retired, Philip E. Culbertson commented extensively on the book as it took shape. I lost count of the number of times he sent me ideas but never forgot what he said. Dr. John E. Naugle, another high-ranking NASA civil servant now retired, offered detailed comments as well and prevented me from committing many errors of fact and judgment. Dr. Walter Haeussermann, Karl L. Heimburg, Dr. George F. Pieper, Oran W. Nicks, and Dr. Rocco A. Petrone also reviewed and commented upon the manuscript in its various forms. Their comments were likewise invaluable. I am especially indebted to other officials of the National Aeronautics and Space Administration who took the time to meet with me and discuss their work with NASA and the organizations from which it came. They were all helpful and thoughtful as I probed their experiences for insights into NASA's organizational culture. So were the people from outside NASA whom I contacted and interviewed, people who had never worked as NASA civil servants but whose familiarity with its operations lent perspective to what I was hearing from the inside. Employees in the Personnel Analysis and Evaluation Division at NASA headquarters helped me organize the survey of NASA employees that I conducted in 1988 and locate data on the NASA work force. The study could not have been completed without their help.

At the American University in Washington, D.C., where I have worked for many years, numerous persons made vital contributions to this work. My graduate assistant for two years, John M. Low, a student in our doctoral program, processed survey data, located details, and commented frequently on the direction that the manuscript took. His knowledge of NASA activities assisted me greatly. Other graduate students who labored on this study were Ari J. Sky and Lucinda K. McKinney. My dean, Cornelius M. Kerwin, provided intellectual and financial support. The willingness of universities to pay their faculty to go off and write is a wonderful gift. Doctoral students listened attentively to the progress of my work, too frequently I fear, but invaluably for me. Elizabeth W. Lister handled administrative details with care and understanding when I grew impatient with them. Jennifer A. Drenning typed changes into the manuscript. Dr. Katherine Farquhar guided me deeper into the literature on organizational culture than I had previously ever been.

Many people commented on the development of the study or provided a forum for my ideas. Dr. Todd R. LaPorte gave important guidance as the study got under way. Patrick B. Nolan of the Center for the History of Business, Technology, and Society at the Hagley Museum and Library in Wilmington, Delaware, invited me to deliver a lecture as my research findings began to jell. Dr. Janice M. Beyer, Dr. Maurice Klein, and Dr. John M. Logsdon commented on the paper and article that emerged from that lecture. Dr. David Rosenbloom, a colleague and editor in chief of the *Public Administration Review*, commented extensively on an article based on another aspect of this research.

I have tried to let the people who have worked for NASA speak for themselves, without judging the merits of what they said or altering their thoughts to fit a theme. This is the culture as they described it. In boiling down their experiences from thousands of pages of transcripts and documents, I hope that I have not made their culture seem less rich than it really is.

Inside NASA

Introduction:
NASA's Organizational Culture

For a generation of Americans, the space program has been one of the few memorable things the government got right. —Richard Corrigan, 1986

Popular wisdom holds that government organizations inevitably decline. They start out with great enthusiasm, but eventually their capabilities wind down. They grow excessively bureaucratic. Their employees resist new ideas. Overall performance deteriorates.[1] Skepticism about the capabilities of large government agencies is an important part of American political life. Does this belief stand up to scrutiny, or is it just part of the natural antagonism toward big government in a democratic society?

The people who ran the National Aeronautics and Space Administration (NASA) when the agency was young developed methods of doing business that allowed them to carry out extraordinarily difficult tasks. NASA civil servants discovered ways to circumvent bureaucratic restrictions and avoid failure when it threatened to occur. They adopted an organizational philosophy suited to the scientific and technological missions they were asked to perform. NASA acquired a reputation as a high-performance government organization.

As NASA grew older, it changed. Beliefs about how the agency should be run persisted, but no longer did those beliefs elicit the behavior that characterized the early years. NASA grew more bureaucratic. It became more concerned with maintaining its survival. In the eyes of some people, its performance declined. The onset of maturity changed NASA. The agency that embarked upon the 1990s was a different one than the NASA that sent astronauts to the Moon. The NASA story helps reveal forces that work to mollify the capabilities of high-performance organizations in the public sector. Unlike business firms, whose fundamental outlooks tend to persist over long periods of time, NASA's organizational culture blossomed and lost strength within just thirty years. The NASA experience suggests that high-performance cultures within the public sector are inherently unstable, given the conditions with which they must deal.

NASA appeared as a new organization on the governmental scene in 1958. Congress sent the legislation creating NASA to President Dwight D. Eisenhower less than one year after the Soviet Union launched *Sputnik 1*, the first earth-orbiting satellite made by humans, on October 4, 1957. The authorizing legislation, which President Eisenhower signed on July 29, 1958, concentrated responsibility for civilian space and aeronautical activities in a new government agency. Responding to what most viewed as a cold war technological challenge, NASA employees and their contractors completed a series of programs that firmly established the United States as the leading space power among the nations of the earth and NASA as a performance leader among government bureaus.[2] Between 1961 and 1975, NASA officials completed the Mercury, Gemini, Apollo, Skylab, and Apollo-Soyuz programs—a total of thirty-one consecutive expeditions into space—without losing a single astronaut on a space flight mission.

NASA engineers perfected the giant Saturn V rocket, which could lift a 285,000-pound (130,000-kilogram) payload into low earth orbit. In 1969 NASA civil servants successfully landed the first humans on the Moon and returned them safely to Earth. During the 1960s, NASA scientists and engineers launched the first weather and communication satellites, which helped to revolutionize human perspectives of the Earth. They sent satellites above the atmosphere to study the Sun and stars and map the newly discovered Van Allen Radiation Belts and Earth's magnetosphere. They launched robots that landed softly on the surface of the Moon and analyzed samples of the lunar soil. In 1972 they launched *Pioneer 10*, the first spacecraft to fly through the asteroid belt and visit Jupiter. *Pioneer 11* and the Voyager twins followed, collectively flying by Jupiter, Saturn, Uranus, and Neptune. In 1975 NASA launched two Viking spacecraft that landed safely on the planet Mars and searched for evidence of life in Martian soil. NASA accomplishments during those years became synonymous with outstanding performance and high reliability. It became fashionable to judge other government programs by space agency standards. If we can land a man on the Moon and send a spacecraft to Jupiter, went the rhetorical question, why can't we end poverty and hunger on Earth?

Yet, later on, scarcely three decades into the era of space exploration, NASA seemed like a beleaguered agency. During the 1970s NASA employees experienced a series of management problems in the development of the Space Transportation System and the Hubble Space Telescope.[3] New problems appeared when the projects flew in the 1980s and 1990s. On January 28, 1986, the space shuttle *Challenger* exploded during its launch, killing all seven members of the crew and wounding NASA's reputation for outstanding management.[4] Troubles plagued the Galileo probe to Jupiter, launched in 1989. A flaw in the mirror of the Hubble Space Telescope, discovered after the launch on April 24, 1990, compromised the quality of the pictures NASA expected to receive and further damaged the agency's standing. The space shuttle failed to meet its original goal of providing routine, low-cost access to space.[5] These problems set in motion a general decline in the confidence once placed in NASA management, a development amplified each time NASA experienced a new problem or a cost overrun.

NASA, it should be said, experienced plenty of failures during the 1960s. The first six Ranger spacecraft, robot probes designed to take close-up pictures of the Moon, failed to perform their missions. Three astronauts died during a launch simulation at the Kennedy Space Center on January 27, 1967, when a fire broke out in a fully sealed Apollo space capsule on launch pad 34. As in the 1980s, these malfunctions sparked investigations both inside and outside the agency.

The federal establishment, along with the public at large, tended to view the failures of the first decade as aberrations from a general pattern of success. The same people interpreted the failures of the third decade as symptoms of NASA's general decline. A survey of former federal executives taken in 1990 revealed how far professional confidence in the civilian space agency had fallen. The executives rated NASA toward the middle of all federal agencies. In the opinion of the executives, the organization that had once produced the Apollo expeditions to the Moon and the Voyager expeditions to the outer planets now stood alongside the Federal Deposit Insurance Corporation and the Internal Revenue Service—still professionally run but no longer among the federal superstars.[6]

No one doubts that technically complicated missions such as those NASA performs pose special challenges. For a NASA program to work, a large number of factors must fall into place. The technology upon which the mission depends must be well developed, a prior investment that is critical for program success. Technically skilled people must be motivated to join the project work force. Year after year, White House budget directors and Capitol Hill legislators must provide sufficient money, a generally scarce commodity in a government long on promises and short on funds. Since at least some of the tasks that make up a complicated mission will certainly fail, success requires a certain amount of luck and a good measure of flexibility. Finally, success requires that the members of the agency charged with carrying out the mission be capable of planning and organizing it, which in turn necessitates an organizational culture that is suited to the task at hand.

This book provides a history of NASA's original culture—its way of doing things. It identifies the culture that supported the Apollo expeditions to the Moon and the other civilian space programs of the early years. It describes the ways in which that culture changed in the second and third decades of space flight. It examines the NASA experience in light of theories that attempt to explain why government organizations alter their fundamental norms of behavior as they age and ends by suggesting why cultures such as NASA's turn out to be so short-lived.

Like societies as a whole, organizations develop distinctive cultures. An organization's culture consists of assumptions, often unspoken, that members make as they go about the process of doing their work. It consists of beliefs widely shared by members about the ways the organization should operate and its employees behave. It consists of practices, the familiar patterns of behavior that emerge from assumptions and beliefs. Norms, customs, values, and language also make up the culture of the organization. Webster's *New World Dictionary* defines culture as "the ideas, customs, skills, arts, etc. of a given people in a given period." Ott states that an organizational culture "is made up of such things as values, beliefs, assumptions, perceptions, behavioral norms, artifacts, and patterns of behavior."[7] Altogether, the culture defines "the way we do things here," the manner in which members expect the organization to operate.

Organizational cultures serve many functions. They guide employee behavior. They shape official decisions. When deeply held, cultural norms influence behavior and decisions in ways that official procedures rarely approximate. As such, an understanding of cultural norms becomes an indispensable tool for predicting the reaction of employees to a new policy or a changing situation. Some of the most insightful studies of government operations deal with the way in which employees in public organizations approach their work.[8] Such insights are more useful for understanding the behavior of people inside organizations than is knowledge about formal structure, official policies, or the objectives the organization consciously pursues. Executives who seek to change the direction of organizations ignore culture at their peril. For people who study such things, organizational culture is a powerful means for predicting how members behave.

Strong cultures help to unify organizations that might otherwise fall apart. Cultural norms lend them a greater degree of cohesion than organizational politics would otherwise allow. Large government organizations, as a rule, have a tendency to come apart. Employees may be separated by geography; they must respond to diverse clientele and conflicting objectives. They develop loyalties to their own institutions. They gravitate toward their own perceptions of their own responsibilities. Altogether, this creates a form of organizational entropy that favors disorder over common orientation.

In his classic study of the U.S. Forest Service, Kaufman explains how a common culture unifies forest rangers otherwise divided by distance and local responsibilities. The size and complexity of the forest ranger's job creates tendencies toward fragmentation. The profession of forestry, at the time Kaufman wrote the book in 1960, created a common culture. Although he did not use the term *culture* to describe it, Kaufman recognized that "much that happens to a professional forester in the Forest Service . . . tends to tighten the links binding him to the organization. His experiences and his environment gradually infuse into him a view of the world and a hierarchy of preferences coinciding with those of his colleagues." The result, Kaufman observed, was "voluntary conformity" with overall Forest Service policy "in spite of the awesome tendencies toward fragmentation to which that organization is heir."[9] Kaufman saw how this "view of the world" became a powerful force for ensuring predictability in the activities of field employees, a force more powerful than budgets, procedure manuals, and other official control mechanisms.

In the early 1980s, as the culture perspective gained respect, writers suggested that culture could play another important function as well. The right kind of culture might strengthen organizational performance. Although all cultures are to a certain extent unique, successful or innovative organizations were thought to possess certain cultural attributes in common. This proposition was advanced considerably by efforts to explain the exceptional performance of Japanese business firms. Authors such as William Ouchi traced the Japanese economic "miracle" to the rise of distinctive corporate cultures different from those to be found in American business firms. Ouchi suggested that American executives, without trying to copy foreign traditions, could learn from Japanese practices such as long-term employment, slow evaluation and promotion, investment in training, and norms of collective responsibility.[10]

The culture-performance nexus received widespread attention following the publication in 1982 of Thomas Peters and Robert Waterman's best-selling book *In Search of Excellence.* Peters and Waterman traced the success of American high performers like Walt Disney Productions and IBM to the development of a distinctive corporate culture. Excellently performing firms, the authors suggested, possessed corporate cultures that emphasized eight key tenets, including a hands-on orientation, close concern for the needs of the customer, a lean central staff, and a bias for action.[11] Studies of other modern firms, such as American Telephone and Telegraph Company and Chrysler Corporation, lent further credibility to the link between culture and performance.[12]

A research project focused on public sector organizations further amplified the importance of distinctive norms in supporting standards of excellence. Members of the High Reliability Organization Project at the University of California, Berkeley, analyzed air traffic control systems, power grids, and naval air operations at sea—all government operations that are able to achieve very low error rates and avoid catastrophic failures. These government organizations, the researchers observed, developed operating philosophies quite different from the more common trial-and-error bureaucracies.[13]

There is a certain amount of circular reasoning in the proposition that successful organizations possess successful cultures. Almost by definition, the culture of a successful organization must be successful. How could a successful organization be based on an unsuccessful culture? Peters and Waterman attacked this problem by identifying common cultural elements among many successful firms in the expectation that what is common to many must contribute to success, although they did not deal with the disturbing possibility that a failing organization might also possess the elements of a successful culture.[14] In NASA's case, the temporal dimension helps to sort out the problem of circularity. Much of NASA's early culture existed before NASA was created, in the predecessor organizations that came together to form the civilian space program. Since those organizations existed prior to the NASA space flight program, it is illogical to presume that NASA's cultural orientation arose as a consequence of its accomplishments.

While scholars have been able to establish an association between culture and performance, they have not been able with convincing clarity to explain cause and effect. No major theory of organizational culture, accompanied by predictions and supporting evidence, has been produced to show exactly why culture affects performance. Neither have scholars been able to show exactly how much of a difference culture makes, as compared to other factors that bear on the performance of an organization, such as proprietary technology or sufficient funds.

In part this inability to gauge cause and effect arises from the aversion many culture analysts show toward conventional social science methods. Social science research methods, with test and control groups and longitudinal analysis, provide useful tools for probing cause and effect. These research methods are not, according to many cultural analysts, useful for revealing norms often shielded by the tacitness veiling beliefs and assumptions. As such, the study of culture remains what J. Steven Ott characterizes as a "perspective" on organizational life rather than a general theory.[15] It is, nonetheless, a powerful perspective.

The effort to link performance to culture is but one of the challenges that observers of the phenomenon face. The identification of cultural norms is a demanding task. Few organizations expose their cultural tenets by writing them down. They may publish general statements of philosophy, but they rarely commit deeply held assumptions to paper in sufficient detail to be of much use to scholars. In some organizations, particularly within government, organizational cultures can be at war with themselves. Many agencies incorporate conflicting cultures. This often results from the practice of amalgamating separate institutions, each with its own distinct culture, into governmental conglomerates. Institutions so created tend to be confederations of cultures rather than one culture fit to a common mold. NASA fits this description particularly well. During its creation, the people who assembled NASA brought together many distinct ways of thinking. They took engineers and scientists from aeronautical research laboratories and put them together with German rocketeers enlisted to develop NASA's large rocket program. Over these groups they imposed officers and industrialists from the U.S. Air Force Ballistic Missile Program. The people assigned to work on what NASA officials called the manned space flight program did not share the perspective of people who worked on unmanned satellites and robotic probes. NASA's research centers, which were government laboratories, operated differently than NASA centers set up to manage space operations. Each group developed its own distinctive assumptions, beliefs, and practices.

Out of this confederation of subcultures within NASA emerged both disagreements and common beliefs. The importance of testing, for example, was a belief shared widely within NASA by German rocketeers and American aeronautical engineers alike. It thus became part of the common culture. Questions about how to test—whether to test incrementally, one element at a time, or to test everything together (all-up)—deeply divided NASA employees. Factions representing different subcultures fought fiercely over issues like these. In some cases, one subculture prevailed. In other cases, differences remained.

Identifying cultural norms is difficult even in the absence of competing subcultures. Some employees are not consciously aware of the cultural assumptions they hold. Among the people who study cultures, a substantial number believe that tenets cannot be discovered through social science techniques. Some go so far as to suggest that the very act of asking an official to describe his or her culture will distort important beliefs. Such researchers favor indirect observations and archive searches to sort out unspoken assumptions.[16]

This study made use of a number of methods to identify and verify NASA's cultural norms. The research began with indirect observation and review of secondary sources describing various NASA programs. An interview program followed. In-depth interviews were conducted with people who led NASA in the 1960s and with their counterparts from the 1980s.[17] Although these people were informed that they were part of a culture study, they were not directly asked to describe NASA's underlying norms. Instead they were encouraged to talk about their work, the things that they valued most about the agency, their experiences with the space flight program, and other subjects drawn from an interview guide (see the Essay on Sources for further detail).

To verify the pervasiveness of beliefs identified during the interview program, a survey questionnaire (referred to as the NASA culture survey) was distributed in 1988 to a random sample of eight hundred NASA scientists, engineers, and administrators. The survey sought to determine whether members of the agency at large shared the beliefs expressed by the people who were interviewed. A small group of outside contractors and project scientists also participated in a check on the interview program. These people had worked with NASA but had never been on the agency payroll. They were interviewed to see how outsiders viewed NASA's cultural norms.

Agency budget and work force statistics provided objective evidence to test the beliefs and assumptions expressed by NASA employees. Agency histories and archives provided evidence to probe the reality behind stories that people told. These sources were consulted with the understanding that an assumption or legend need not always be objectively true in order to be a motivator of behavior.

Finally, with the aid of graduate assistants, NASA archives were examined to compare comments made some twenty years after the events of the first decade with observations made at that time. This helped to identify those perceptions that remained constant through time and to avoid the warping effects of nostalgia. A more complete description of the research methodology is contained at the end of the book in the Essay on Sources.

From this research a pattern gradually emerged. The practices and assumptions making up NASA's early culture appeared. So did the sense, widely held among people familiar with the civilian space program, that NASA's underlying way of doing business had somehow changed. The following chapters describe the elements of NASA's early culture, identify the ways in which people believe that this culture lost vitality, and offer an explanation of why that transformation took place.

1 ▲ Building Blocks

Department heads seldom start with a clean slate. . . .
There are likely to be daily reminders that they are
merely temporary custodians and spokespersons for
organizations with distinct and multidimensional
personalities and deeply ingrained cultures and
subcultures. —Harold Seidman, 1970

The building blocks for NASA's overall culture came from many places. The government research laboratories of the National Advisory Committee for Aeronautics changed their names as they transformed themselves into NASA but not their well-established cultures. Workers from the Army Ballistic Missile Agency, led by a group of German expatriates under Wernher von Braun, brought their philosophy of rocket engineering to the new space program. The U.S. Navy contributed personnel from its own Naval Research Laboratory. The U.S. Air Force contributed the organizational philosophy of its intercontinental ballistic missile program, along with a few of the missiles themselves. NASA acquired the Jet Propulsion Laboratory, a contract operation run by the California Institute of Technology, a leading American university. NASA executives created new field centers for spacecraft design, space science, and rocket launching, but did so with people already set in their ways. Diverse origins created a cultural confederation, with differences often exceeding common orientations.

The Research Laboratories

When high government officials require the assistance of experts, they often turn to universities or privately run research laboratories where those experts work. To build the American space program, public officials turned to the government itself. When the space race began, the U.S. government already had a forty-year-old institution devoted to the investigation of flight.

In 1915, with the war in Europe providing a catalyst, the U.S.

Congress established a National Advisory Committee for Aeronautics (NACA) "to supervise and direct the scientific study of the problems of flight, with a view to their practical solution."[1] The fledgling committee commissioned studies of propellers, encouraged the development of aircraft engines, and in 1920 formally opened a small research laboratory at an Army air base along the Chesapeake Bay a few miles north of the tidewater town of Hampton, Virginia. The Langley Memorial Aeronautical Laboratory was led by engineers who acquired a reputation in this small Virginia town as "NACA nuts," a reputation founded as much on their odd demeanor as their technical expertise. They stood apart from the local citizenry, a distinctiveness that led many to assert that they were also different from the average government employee. They did not work for the federal bureaucracy, many said, they worked for the *N-A-C-A*. They took pride in the assertion that, as experts working in a research laboratory (albeit one funded by tax revenues), they were obliged to pursue technically correct solutions to flight engineering problems. They were not at all reluctant to assert their sense of intellectual superiority when confronted with what they considered to be a compromise in technology.[2]

In 1939, the year in which the second world war to be fought with airplanes began, Congress authorized a new aeronautical laboratory to complement the work of the Langley facility: the Ames Aeronautical Laboratory at Moffett Field in California, at the southern end of San Francisco Bay. Staffed by a nucleus of experts from the Langley laboratory, the Ames facility took on the culture of the "engineer in charge," a technical oasis in a government of bureaucrats and politicians. As at the Langley facility, the Ames staff pursued investigations into the science of aerodynamics, with a particular concentration on high-speed flight.[3]

One year later, Congress created a third laboratory in Cleveland, Ohio, to conduct aircraft engine research. It did so in response both to news that British and German engineers had developed aircraft engines capable of operating at higher speeds and altitudes and to a secret report of European plans to develop a radically new propulsion technology, the turbojet engine. Engine research had not been a major priority within the NACA, that function having been left to private industry.[4] The third laboratory developed a culture with a slightly different focus than the model-building aerodynamic emphases at Ames and Langley, although it retained the commitment to fundamental research. The Lewis Flight Propulsion Laboratory, as it became known, was located on a site bordering the Cleveland municipal airport.

Together, the Langley, Ames, and Lewis laboratories provided the United States with its premier institutions for flight research during the development of the airplane. NACA technical reports were widely used by university students taking courses in aeronautical engineering. The field operations maintained the culture of the research laboratory, of the engineer in charge, of the triumph of technology and scientific inquiry for problem solving. They maintained a tradition of technical detachment. The laboratories were geographically isolated, distant from the politics of Washington, D.C. They did not have to report to a large Washington bureaucracy. The advisory committee to which the laboratory directors reported in Washington, in step with the concept of limited government prevalent at that time, kept the NACA headquarters staff small. Until the space age began, the Washington staff of the NACA never exceeded two hundred employees. Even the demeanor of NACA employees set them apart, their precise, demanding approach to problems making them appear socially detached.

NACA civil servants also drew on another ethos, one epitomized by the work conducted at an unpretentious field station set up in 1946 by a group of Langley engineers and technicians to study the dynamics of high-speed flight. These employees organized the High Speed Flight Research Station on a barren portion of the Mojave desert in southern California, adjoining a remote U.S. Army Air Force test site near a large, dry, lake bed. The military had arrived a decade earlier. Working with the Army Air Force and the Bell Aerosystems Company, Langley employees tested the Bell *XS-1*, a cigar-shaped experimental aircraft propelled by a rocket engine that consumed liquid oxygen and diluted alcohol.

The work proved dangerous. Geoffrey de Havilland, one of Britain's best test pilots, lost his life in 1946 when his experimental aircraft broke up during a high-speed dive over the Thames estuary. A Bell Company test pilot, Jack Woolams, died that same year. NACA and later NASA employees crashed twelve aircraft and lost four test pilots to flight accidents in slightly more than two decades at what became known as the Hugh L. Dryden Flight Research Center.[5]

The NACA high-speed flight project pushed the edge of the performance envelope, using technology to do what humans had not done before. On October 14, 1947, an Air Force pilot working with the NACA team, Captain Charles E. Yeager, pushed the Bell *XS-1* through the sound barrier. Twelve years later, NASA recruited its first astronauts from the exclusive fraternity of test pilots. Test pilots understood the dangers of exploration and provided public witness to the fundamental belief that flight test programs—whether in the air or in space—existed to break performance boundaries. Their presence reinforced the basic assumption that flight performance problems would respond only to technical solutions. In a flight test program, the price of a compromised engineering judgment could be awfully high.

By October 1957, when the beeping of the Soviet *Sputnik 1* earth satellite inaugurated the age of space exploration, NACA employees had already conducted a good deal of research on the technology of space flight. Engineers at the Langley Pilotless Aircraft Research Division had tested the aerodynamics of space flight by firing rocket models out over the Atlantic Ocean from a modest launch facility at Wallops Island, Virginia. Experts at the Ames laboratory had completed work on the blunt-body theory that would guide the design of spacecraft traveling to and from the earth. A group at the Lewis laboratory had developed a liquid-hydrogen-fueled rocket, a high-energy technology that would make possible flights to the Moon. Workers at the High Speed Flight Station in the Mojave desert were making preparations for the test flights of the *X-15*, an aircraft that could fly to 350,000 feet. NACA Director Hugh L. Dryden argued that the forty-three-year-old agency "has been engaged increasingly in research applicable to the problems of space flight" and urged that the government give the NACA a major role in any new national space program.[6] On October 1, 1958, the NACA research laboratories became NASA, changing their nameplates but not their well-established orientation.

In spite of their previous work, none of the NACA installations had actually developed a vehicle that could reach outer space. To do that the old NACA research laboratories needed the cooperation of groups that would add two more layers of complexity to the NASA culture confederacy.

The Rocket Engineers

During the years that followed World War II, large rockets were military devices. NACA employees conducted research on propulsion and the aerodynamics of missiles and their warheads, but they did not develop their own large rocket program. That task belonged to the military services, each of which made a contribution to the incipient space program through its development of military hardware.

Not until the mid-1950s did the U.S. Air Force undertake a major effort to mass produce long-range rockets for military use. This delay had a profound effect on the development of the civilian space program.

In 1947 the Air Force canceled its long-range missile program with the Convair Corporation, redirecting funds into the effort to build long-range bombers capable of penetrating Soviet air space. In 1953 the Soviet Union exploded its first thermonuclear bomb, only one year after U.S. officials had done the same. Nuclear scientists and missile experts told U.S. officials that it would be possible within a few years to build hydrogen bombs small enough to fit inside the nose cone of a long-range rocket; intelligence data indicated that the Soviets were already working on such a program.

Fearing that the Soviet Union would attain nuclear superiority over the United States, top government officials directed the Air Force to undertake a crash program to deploy long-range rockets capable of firing nuclear warheads into the Soviet Union. The Air Force missile program had the scope and urgency of the Manhattan Project, the effort undertaken fifteen years earlier to develop the atom bomb. In part because they needed to develop the rockets so rapidly, Air Force officers rejected the army arsenal concept, wherein the development and initial construction of the rockets could have taken place at government facilities. Instead, they adopted a system of parallel contracting, whereby hundreds of privately owned companies simultaneously designed and fabricated program components. The Air Force even relied upon contractors to help coordinate other contractors, performing the complicated systems engineering function and providing technical direction for the overall program.[7]

Although the Air Force developed an interest in having its own space program (and even established its own astronaut corps during the 1960s in anticipation of building its own space station), preoccupation with the ICBM program during the 1950s removed the Air Force from the competition to launch the first U.S. earth satellite. That race fell to the Navy and Army, which aggressively touted the merits of their respective programs. In 1955, a special advisory group selected the Navy proposal, ostensibly on technical merit.[8] The Deputy Secretary of Defense authorized the Navy Department, through its Naval Research Laboratory, to begin work on a small scientific satellite to be launched during the 1957–58 International Geophysical Year. The laboratory, an in-house government facility located along the Potomac River in southeast Washington, D.C., was run along much the same lines as the field installations of the NACA. It too would make an important contribution to NASA's organizational culture.

The history of Project Vanguard, as the Navy's earth satellite program was called, would have been comical were it not for the anxiety generated by the successful orbiting of the Soviet Union's *Sputnik 1* on October 4, 1957. A nation capable of propelling satellites into space might also pull ahead in the technology supporting the arms race. It surely gained a psychological advantage in the contest for world opinion. *Sputnik 1* and its successor, *Sputnik 2*, launched less than a month later, provided a frightening demonstration of accomplished technology. The United States planned to reassert its technical prowess using a thin three-stage rocket. The first two stages of the rocket were based on the technology of sounding rockets, small vehicles used to shoot instruments into the upper atmosphere. On December 6, 1957, the Vanguard rocket rose three feet from its launch platform, paused, and began to fall. The lower stage exploded in flames, the upper stages broke apart, and the three-pound satellite rolled away, beeping pointlessly.

Less than two months later the United States successfully orbited *Explorer 1*, a small scientific satellite. The Juno I rocket that carried the satellite most of the way into orbit had been developed at the Army Ballistic Missile Agency (ABMA) in Huntsville, Alabama. German expatriates (by then U.S. citizens) and their leader, Wernher von Braun, a space exploration visionary whose exploits theretofore had been confined to the development of military missiles, directed the team.

The German rocket team had assembled itself during the late 1930s at the large military research facility at Peenemünde, on the Baltic Sea. Carefully refining their technology, they managed to produce their first large, long-range rocket by 1943, which was used to deliver conventional explosives to targets on the Continent and in Britain. As the war ended, members of the German rocket team, including von Braun, fled to Bavaria so as to be captured by the Americans instead of the advancing Soviet army. Eventually, some 125 German rocketeers reached the United States, where they hoped to continue their rocket development work under the auspices of the U.S. Army. The Army sent them to El Paso, Texas, where they countinued testing their V-2 rockets at the nearby White Sands Proving Grounds in New Mexico. In 1950 the Army moved them to the Redstone Arsenal at Huntsville, Alabama, along with some eight hundred supporting employees. In 1960 the group moved into NASA with 4,500 employees—and the Germans still firmly in control.[9]

The rocket team brought a specific culture to NASA. Members of the team were meticulous in their work and believed strongly in the need to give exacting attention to small details. They believed in proceeding incrementally, step by step, in the development of rocketry.

The Germans believed in controlling all aspects of a project and doing as much work as possible with their own staff. That meant control of not only the technical labs where the basic work of guidance and propulsion was done, but the fabrication shops as well. The fact that the military sent the German rocket team to an army arsenal, a place for manufacturing munitions, helped perpetuate this tradition.

Each of the military service programs made an important contribution to NASA's emerging organizational culture. The Army rocket program of the ABMA joined NASA on July 1, 1960, bringing with it the habits of the German rocket team. NASA called it the George C. Marshall Space Flight Center, dedicated to the Army officer and Secretary of State who helped rebuild Europe after World War II with the Marshall Plan. NASA also acquired the staff and facilities of the Jet Propulsion Laboratory in Pasadena, California, an Army contract operation that had worked with the ABMA to develop the first U.S. satellite, *Explorer 1*. From the Navy, NASA acquired the staff of the Vanguard satellite program at the Naval Research Laboratory. Once the manned space program got under way, the Air Force contributed two of its intercontinental ballistic missiles, the Atlas and Titan, which were used to propel the Mercury and Gemini astronauts into orbit. More important, the Air Force contributed the management culture of its large rocket program, which was absorbed into NASA once the lunar expedition started through the judicious transfer of key Air Force and contract personnel.

Human Space Flight

NASA officials wanted to conduct their own operations in space. With the acquisition of the ABMA, they acquired the capability to develop rockets large enough to send humans to the Moon. What they lacked, at that time, was a facility devoted exclusively to spacecraft. They also needed their own launch center. The development of those capabilities added new layers of complexity to NASA's confederation of cultures.

At first, NASA officials concentrated responsibility for the human space flight program at the Langley Research Center, the forty-year-old aeronautical laboratory transformed into a NASA field center. The original *Mercury 7* astronauts reported for duty at the Langley Center. A special Space Task Group, led by members of the old Pilotless Aircraft Research Division, directed work on the Mercury spacecraft. Through ten years of model-testing activities, the members of the Langley Pilotless Aircraft Research Division had acquired a basic understanding of the aerodynamics and thermodynamics of reentry, which they promptly applied to the design of the cone-shaped Mercury capsule.

NASA officials assumed from the beginning that the Space Task Group would move to a new NASA field center, one devoted primarily to space operations. This was a crucial assumption in the development of NASA's early culture. The Langley Center, like other NACA laboratories, continued to conduct research and produce technical reports. Its employees developed a few space expeditions, notably the Lunar Orbiter and the Viking lander projects, but it remained fundamentally a research facility. Space operations had to be separated from the Langley facility, so as not to overwhelm its research orientation.

The Space Task Group moved out in 1962, relocating on salt grass pasturelands to the south of Houston, Texas. The Manned Spacecraft Center (renamed the Lyndon B. Johnson Space Center in 1973) became NASA's operational center for human space flight. It acquired the training facilities for NASA's astronaut corps. It housed the Mission Control Center, from which NASA employees could direct space flight operations. The members of the Space Task Group brought from the Langley center the engineering talent needed to design ships that could fly through space, including the Apollo space capsule, the lunar landing module, and eventually the space shuttle orbiter.

Because it grew out of the Langley Pilotless Aircraft Research Division, the Houston center retained much of the cultural orientation of the old NACA. Its employees believed thoroughly in the importance of research and testing. They insisted on seeking technical solutions to space flight problems, with a minimum of outside interference. Associated as they were with the test pilots in the astronaut corps, they adopted the ethic of taking risks to push performance frontiers. In only one respect did the Manned Spacecraft Center depart significantly from the Langley research culture: it relied significantly upon contractors for spacecraft fabrication and technical assistance.

Cultural tension built up between the employees of the Manned Spacecraft Center and what rapidly became NASA's other major center for human space flight. Von Braun's rocket team, which joined NASA in 1960, wanted to build more than rockets. In 1952 von Braun had proposed the construction of an earth orbiting space station, serviced by a winged space shuttle that could fly between the station and the earth.[10] Marshall Center employees eventually received the assignment to build *Skylab*, NASA's first orbital workshop, the propulsion system for the space shuttle, and the modules for the space station *Freedom*.

Competition for assignments between the Marshall and Johnson centers accentuated preexisting cultural differences.

Although Marshall Center and Johnson Center employees shared many cultural norms, such as their belief in research and testing, they derived those norms in different ways. Johnson employees perpetuated the flight research traditions of the NACA, which had played a leading role in the development of American aeronautics. They were proud of the U.S. capabilities in that field. The German rocket team traced their heritage to the Army arsenal system and, before that, to Peenemünde. Its members possessed a tradition of deep control over all elements of their projects, relying upon their own employees as much as possible. Twenty-six years after the two institutions came together within NASA, when the original cast of characters had long since departed, a presidential commission investigating the explosion of the space shuttle *Challenger* pointed out the continuing tendency of the two centers to diverge. The commission criticized the propensity of Marshall Center managers to solve problems internally without telling other program managers about their difficulties, especially those at the Johnson Space Center to whom they officially reported.[11]

To complete their human space flight capability, NASA officials wanted to construct their own launch facilities. Mercury and Gemini astronauts, as well as NASA's first scientific satellite, lifted off from launch facilities under the jurisdiction of the U.S. Air Force at Cape Canaveral, Florida. Through interagency negotiations, NASA officials sought more control over their own satellite and spacecraft launches. Nonetheless, they remained guests on a facility that the Air Force, as the "host agency," still controlled.[12]

Contractors began work on an all-NASA launch facility on Merritt Island, to the northwest of the Air Force station, in January 1963. The development of the John F. Kennedy Space Center, as it became known, fell largely to the Marshall space flight team. German rocketeers had been firing rockets from Cape Canaveral since 1950. They had stationed their own employees at the cape, who in turn formed the nucleus of the Launch Operations Center that NASA set up for its early rocket launches. Kurt Debus, von Braun's chief technical assistant and a member of the Peenemünde group, headed the Launch Operations Center. A U.S. Army officer who had worked closely with the von Braun team during the 1950s, Rocco Petrone, oversaw construction of the all-NASA "moonport." NASA conducted the first launch from its new Merritt Island facility on November 9, 1967, an unmanned test flight of the Saturn V rocket and Apollo spacecraft.[13]

Both the Manned Spacecraft Center in Houston and the Kennedy Space Center in Florida grew out of existing institutions with well-established technical cultures. The Houston center was a product of the Langley NACA aeronautics culture; the Kennedy launch facilities were organized by the von Braun rocket team. While the activities of the two new centers imposed special requirements, particularly an increased reliance upon contractors, their underlying practices and philosophy were easily recognizable to someone familiar with the parent institutions.

The Science Centers

Originally, NASA officials planned to locate both the manned space flight program and the unmanned satellite program at the same center, a new installation some twelve miles northeast of downtown Washington adjacent to the suburban community of Greenbelt, Maryland.[14] NASA executives wanted to use the Vanguard group from the Naval Research Laboratory in southeast Washington to form the nucleus for that part of the center responsible for unmanned satellites. The Space Task Group and the Mercury astronauts from the NACA Langley Center in southeast Virginia would form the nucleus of the human space flight effort. Agency officials formally established what became the Goddard Space Flight Center within months of NASA's founding, even though the groups that made up the new center still worked elsewhere.

The rapid growth of the manned space flight program, occasioned by President Kennedy's 1961 decision to dispatch an expedition to the Moon, prompted NASA officials to reconsider this plan. They decided to relocate the manned program at a site along a waterway where spacecraft could be barged to their launching site.[15] In September 1961, NASA officials formally announced that the new Manned Spacecraft Center would be located south of Houston. This left the Goddard Space Flight Center with a responsibility that did not extend beyond unmanned satellites and robotic flight. The two hundred employees from the Naval Research Laboratory who formed the nucleus of the Goddard Center had little contact with the groups being organized to send people into space.

The division of responsibilities contributed to a continuing cultural schism within NASA. NACA employees responsible for manned (human) spacecraft went to what became the Johnson Space Center near Houston. Much of the unmanned (robotic) space flight program went to Goddard. While the cultural traditions of the Naval Research Laboratory were in many ways similar to those at Langley, the NRL was a Navy Facility quite distinct from the NACA. It was a separate organization with its own history. Although most large NASA undertakings combine elements of manned and unmanned flight—robots preceded humans to the Moon, for example—each group developed its own institutions and clientele. The schism might not have been as pronounced had NASA officials insisted on their original plans to merge satellite and human space flight operations at one new center.

Goddard employees developed most of the Explorer satellites that studied such phenomena as solar and cosmic radiation and the earth's magnetic field. They developed the first Orbiting Astronomical Observatories, the precursors for large observatories like the Hubble Space Telescope. They acquired an early expertise in the construction of automated orbiting platforms, such as the Orbiting Geophysical Observatories, on which twenty or more experiments could be mounted, and the Orbiting Solar Observatories. They acquired a reputation as a science center. All of the NASA centers conducted scientific research, but the ties that Goddard Center employees developed with scientists who flew their instruments on board the Goddard satellites gave the installation its reputation for being especially close to the science community, which showed more interest in robotic satellites and probes than in manned spacecraft.[16]

NASA officials established one other center for automated space flight. Practically at the instant of its founding, NASA acquired the Jet Propulsion Laboratory. The JPL had a curious history. It was originally set up as a research laboratory for aeronautics and rocketry under the California Institute of Technology in Pasadena, California, one of the nation's best-regarded research universities. After 1940 it developed into a contract operation for the U.S. Army. In collaboration with the von Braun group at the Army Ballistic Missile Agency, JPL employees planned the launch of *Explorer 1*, the first U.S. satellite. JPL employees developed the satellite and the fourth stage of the rocket that inserted it into orbit. James Van Allen of the University of Iowa designed the satellite experiment that discovered the radiation belts that were to bear his name, affirming JPL's tradition of working closely with scientists outside the laboratory.

Within NASA, JPL employees established their reputation as scientists who could send out probes to explore the solar system. To precede the astronauts to the Moon, they developed the Ranger and Surveyor probes. They developed the Mariner probes that visited Venus, Mercury, and Mars. They developed the Voyager twins that flew by Jupiter, Saturn, Uranus, and Neptune. They built the Deep Space Network, through which employees sitting at a control center at the JPL could communicate with probes travelling to the planets and their moons. As with the Goddard Center employees, JPL officials maintained a preference for unmanned flight.

Although they conducted some of NASA's most spectacular missions, and participated as coequals on internal agency committees, JPL employees never officially became part of the U.S. government. They remained a contract operation run for NASA by the California Institute of Technology and, as such, retained much of the culture of their parent university, with its strong commitment to scientific research.

A Confederation of Cultures

NASA thus developed a confederation of cultures, a collection of institutions, each with its own history and traditions. NASA acquired three research centers from the NACA—Langley, Ames, and Lewis—with their strong laboratory cultures and their orientation toward aeronautical research. It retained the flight test facilities at Edwards Air Force Base in California (renamed the Hugh L. Dryden Flight Research Center in 1976 and subsumed under the Ames Research Center in 1981). It acquired the Army Ballistic Missile Agency and its German rocket team, which became the Marshall Space Flight Center. It developed two new operational centers for manned space flight, the Johnson Space Center in Houston (originally called the Manned Spacecraft Center) and the Kennedy Space Center in Florida. Each of these centers could trace their early history to the principal organizations out of which NASA was formed, the NACA and the ABMA, respectively. NASA also acquired rockets and managerial expertise from the Air Force Ballistic Missile Program, not in the form of a separate field installation, but as an important component of the new human space flight activity.

NASA officials developed the Goddard Space Flight Center for what became unmanned space flight operations, drawing initially on the traditions of the Naval Research Laboratory. They acquired the Jet Propulsion Laboratory, with its special ties to the California Institute of Technology.

NASA also developed a number of smaller installations: the Michoud Assembly Facility for building rockets outside New Orleans; the Wallops Flight Facility for launching balloons and sounding rockets from a Virginia barrier island; the White Sands Test Facility, part of the

Figure 1   NASA Installations

old missile range used by the U.S. Army to test V-2 rockets after World War II; and a facility for testing large rocket engines in southwest Mississippi that became the John C. Stennis Space Center. From 1964 to 1970, NASA also operated the Electronics Research Center in Cambridge, Massachusetts.

At times, NASA centers developed programs that stretched their basic orientations. Employees at the Ames Research Center developed the Pioneer spacecraft that investigated the atmosphere of Venus and preceded the Voyager twins to Jupiter and Saturn. Langley Center employees developed the Lunar Orbiters that mapped landing zones in anticipation of human flights to the Moon and the Viking landers that set down on the surface of Mars. The inheritors of the German rocket tradition at the Marshall Space Flight Center developed the *Skylab* orbital workshop and bore overall responsibility for the development of the Hubble Space Telescope. The fundamental outlook and responsibilities of the NASA installations, however, remained relatively unchanged.

How is it possible, given this conglomeration of histories, to speak of a single NASA culture? In many ways, the separate centers behave like rival universities, each with its own traditions and interests. Each cradles its own set of contractors, or adopts its own committees of outside advisors, people who are familiar with the way in which the center conducts its business. Each has its own long-range plans and priorities.

This illustrates an important point about organizational cultures. Cultural norms tend to be fairly resilient among the individuals who hold them. The norms bounce back into shape after being stretched or bent. Beliefs held in common throughout the organization resist alteration. Within NASA, however, different individuals treasure different norms. In such cases, dominant norms emerge as a result of the balance of power existing at one particular time. As that balance of power shifts in response to missions and events, the overall ethos can change. Individuals may hold fast to their own beliefs, but the norms for the organization as a whole modulate, based on the relative power of the contributing groups and the situations with which the organization must deal.

Cultural politics of this sort is only one of the factors creating change in complex high-performance organizations like NASA. Other forces, some emanating from the government as a whole, encourage alterations. The following chapters describe the origins and alteration of NASA's way of doing business, beginning with a description of the original NASA culture, moving through the impact of the Apollo buildup, and concluding with the transformation of NASA as the civilian space agency matured.

2 ▲ Root Assumptions

*A good engineer gets stale very fast if he doesn't
keep his hands dirty.*—Wernher von Braun, 1964

When asked to identify the heroes of the American space program,
NASA engineers and scientists interviewed for this study invariably
named their top managers. They rarely mentioned the astronauts. Instead, they recalled the influence of fellow engineers and scientists like
Hugh Dryden, Robert Gilruth, Wernher von Braun, Robert Seamens,
Abe Silverstein, and George Low, people who had risen—often on the
basis of their technical skills—to high administrative posts in the infant
space program.

The engineers- and scientists-turned-managers who shaped the
technical culture of the NASA space program did so using material they
found in the predecessor institutions. Nearly all of these people had
extensive experience in the NACA, the flight test operations at Edwards
Air Force Base, or the ABMA. NASA's first two administrators, Keith
Glennan and James Webb, both political appointees from outside the
agency, picked as their principal deputy Hugh Dryden, the last director
of the NACA. The directors of the eight NASA field installations in
1961 possessed a cumulative total of 185 years of experience with the
predecessor organizations.[1] The directors had strong personalities. They
were committed to the root cultures and they were not tolerant of
actions that threatened to weaken them.

Organizational cultures are often shaped by charismatic leaders
who implant their values in newly forming institutions.[2] In NASA's
case, the leaders drew their root assumptions from government organizations that had developed well-established methods for doing their
work. NASA's early culture emerged from those assumptions, filtered
through the requirements created by the demands of space exploration.

Although NASA's founders drew their values from separate traditions, consensus outweighed division on a number of fundamental issues. The importance of testing as the principal means for verification stood out as one of those issues. NACA engineers and ABMA rocketeers, the two largest groups to form NASA, placed a great deal of faith in the importance of testing.

The NACA test philosophy grew out of its research culture. By the late 1950s, when it was transformed into NASA, the NACA had become one of the world's premier institutions for aeronautical research. The primary products of the NACA were its technical papers, findings from research investigations into the basic scientific principles underlying all types of flight. A technical paper might, for example, deal with the dynamics of airflow on different wing shapes, the result of testing models in laboratory wind tunnels. It was thoroughly scientific and practically always done in-house. The technical papers addressed questions posed by the National Advisory Committee itself, a board of distinguished experts who served part-time; by other government agencies, primarily the military services; by outside organizations such as the aircraft industry; by the NACA staff; and by special technical advisory committees.

The reports contributed significantly to the development of the aeronautical industry and military flight technology in the United States. University students taking courses in aeronautical engineering read NACA reports. "As I was taking aeronautical engineering courses in school," said one NASA executive who saw the NACA become NASA, "I constantly ran into references to the NACA." Upon graduation, he was given a choice between a career in industry and one in government. "The NACA was a totally research-oriented organization," he recalled. "When I thought of the NACA, I thought of research laboratories, which is what I wanted to do." When he thought of airplane companies, he thought "of bull pens of engineers grinding out data," a prospect that did not hold as much appeal.

"I graduated first in my class [in 1958] and I never considered any of the companies," he said, even though the government pay was half of what the companies offered. "It was $4,480," the NACA offer, "which I took, because that was exactly what I wanted to do, and I have never regretted that."[3]

"In 1936 there was no better place you could go than the NACA if you wanted to . . . do research," said another. He had just started work for the Douglas Aircraft Company when a job offer came from the NACA. The Langley Aeronautical Laboratory was 3,000 miles away. "So I wondered, should I take this? My colleagues, the people there, said, 'What are you, stupid? Obviously take it. Jump on it.' If you wanted to do research or development, the NACA was the place."[4]

"You couldn't study aeronautics without knowing about Langley Field," said a NASA center director who joined the NACA in 1937. At that time, the Ames and Lewis research laboratories had not yet been built. NACA research facilities were concentrated at Langley Field. "Most of the data on airfoils, on airplane structures, and on propellers and all those kinds of things came out of the people at Langley. There were technical reports published by the government that you could buy and, of course, the university had copies of them and that's where I read and heard about the NACA. It was my idea that someday I was going to get there and work there. That was my big ambition. And sure enough as soon as I got my master's degree I was off for good old Langley Field."[5]

NACA employees produced more than 16,000 research reports. "The goal of the NACA was to provide the best aeronautical research that could be provided in all fields, whether it be in low-speed aerodynamics, high-speed aerodynamics, gas dynamics, structures, or guidance and control," said another NASA executive who started his career with the NACA. The facilities from which that research emerged were among the finest in the world. NACA engineers could test new theories in a wide variety of wind tunnels, some so large that a whole airplane could fit inside. They built facilities for conducting research on aircraft engines, tracks for testing landing loads, laboratories for testing instruments, and machines for testing structures. "Anybody that had anything [to do] with aeronautical engineering, even in the university, was aware of NACA activities—in particular the NACA technical reports—and knew that the NACA was the prime aeronautical research organization in the United States and probably the world."[6] NACA work was glamorous, sitting as it did on the cutting edge of aircraft flight.

The NACA research centers were relatively small organizations, isolated from the major administrative operations of the federal government. "It was a very fine environment at Langley Field when I came in there [in 1929]," said another NACA engineer who went on to become a NASA center director. Fewer than 200 people were employed by the NACA laboratory at Langley Field in 1929. Headquarters staff in Washington, D.C., numbered twenty one. "Everybody was busy attacking one or another problem of flight."[7] By 1958, when the NACA became NASA, the number of engineers and other professionals on the Langley staff had grown to 1,151, still a modest operation by government standards.[8] "It didn't seem, you know, the idea of a typical government agency," said another one of NASA's leading engineers who started work at the NACA. "It was more like working at a university campus doing research."[9]

Although they were federal employees, few of the NACA employees interviewed for this study thought of themselves as government bureaucrats. They saw themselves as something separate from the bureaucratic caution and political interference that characterized other agencies. "The wonderful part of the whole NACA setup was that it was not a political organization," said one. Organized as an independent agency, the NACA was directed by a committee made up of representatives from the military services, the Smithsonian Institution, the Weather Bureau, the Bureau of Standards, universities, and industry.[10] "We were not a political organization, a bureaucracy in the same sense that NASA is," this employee continued. "The NACA never really operated in a bureaucratic fashion."[11] Since much of the NACA's aeronautical research was done for the military, it acquired a powerful client that could shield it from the political cross fire affecting other civilian agencies.

Given its geographical dispersion, its insulation from politics, its protective clientele, and the fact that it performed research, the NACA developed into an organization in which technical and scientific standards became the dominant criteria for decision making. NACA employees developed a technical culture rather than the more conventional bureaucratic culture that flourishes in large government institutions. The pursuit of truth in any well-run research organization—whether it be a university or a government laboratory—proceeds through testing and verification, not through doctrine or compromise. Research scientists seek to determine the validity of a particular approach by setting up an experiment and studying the results. Someone thinking about the financial or political implications of test results might be upset with the findings, but that would not affect their validity. Only another scientifically accurate test could do that. Test results are test results, no matter how disturbing they may be. They are part of the research scientist's passion for truth, not part of politics, which is viewed as an irrelevant search for compromise.

When the NACA became NASA, the nature of its product changed. NASA managers commissioned the construction of rockets, spacecraft, and satellites. Space flight supplanted aeronautical research as the primary task, even though the old research centers still continued to produce technical reports. Yet while the primary product changed, the basic approach to producing it did not. With the old NACA research centers at the core of the new space flight agency, NASA space flight managers prepared for their tasks through extensive research and testing.

In little more than a year after NASA was formed, agency officials produced an internal long-range plan. The plan set forth the basic philosophy for achieving the exploration goals, in addition to the goals themselves. "During the coming ten years," NASA officials wrote in 1959, "NASA activities will involve extensive programs of engineering development and scientific research." New missions would be approved; new equipment and facilities "often requiring major advances over current engineering practice" would be designed and built. Achieving the necessary levels of reliability "will require extensive testing on the ground and in flight of both vehicles and their payloads." The officials writing the report frankly admitted that this would raise the cost of the space program and "bring about the long lead times associated with space activities."[12]

"There are two general categories of tests," one of the directors of the Apollo spacecraft program wrote in 1970. The first set of tests was performed "on a single prototype device (or on a few devices) to demonstrate that the design is proper and will perform properly in all environments." Tests were performed on each component of the Apollo spacecraft, for example. The escape rocket was tested. The ascent and descent engines were tested. The whole spacecraft was placed in a thermal vacuum chamber to simulate travel in space. The spacecraft was dropped on land and water to simulate descent. The tests gave NASA officials "a tremendous amount of time and experience" upon which they relied when the spacecraft and its astronauts actually flew.[13]

NASA scientists and engineers used prototype tests as a means to analyze components and discover their faults. A spacecraft relay might have to work a thousand times during a mission. "You would take a bunch of them," said one center director, and "cycle them under various conditions, hot and cold and you would just test the hell out of them until they broke down. Then you would fix however it broke, until finally you got something that went beyond the numbers" that project officials thought they had to have.[14]

Such tests were frequently carried out to discover how hardware worked. "You flew things to try them out," said one of NASA's top engineers, "not to prove that you were so smart that your design wouldn't fail to begin with." This was an important element in NASA's test philosophy. "You flew things to find out if they would work, not to prove that we knew how to build them so that they wouldn't fail." Engineers might want to know how a particular shape would heat up as it reentered the atmosphere. "Well, we put some instruments on it and we would fly it and we would find out what the heating was. You made experiments to find things out. Separating a flight test from an experiment never occurred to us." The first parachute tests on the Mercury capsule were designed to find out whether the parachute would actually slow the descent of the capsule, not to prove that the parachute worked. NASA officials conducted seventeen unmanned flight tests of the Mercury capsule to see how it would work before they let John Glenn fly one into orbit around the earth. "If we had needed more, we would have flown more times."[15]

Testing did not stop with prototypes. Once the design work was done, a contractor would fabricate the spacecraft. After the contractor produced the finished product, the second set of tests began. These tests were performed "on each flight item to assure that there are no manufacturing errors and that the item will function as intended." Again, the testing would start with individual components, beginning at the plant where the component was manufactured and again as the component was brought to the assembly site to be installed in the spacecraft. Vibration and thermal tests would be conducted on individual components in order to find worker errors. Factory testing of the whole spacecraft would occur. "First, the wiring is wrung out, and individual subsystems are tested as installed. Then, groups of systems are jointly tested. Finally, the complete spacecraft, with all of its systems functioning, is run in an integrated test." Similar (and at times identical) tests would be run again at the launch site. The spacecraft, with astronauts on board, would be put in an altitude chamber for another complete system check. "The final acceptance test," the official observed, "is the countdown itself."[16]

NASA employees could have accepted the spacecraft from the manufacturer without testing it, based on the certification provided by the manufacturer that all necessary tests had been performed and that the spacecraft met government specifications. This would have been less expensive, at least in the short run. More testing could double the cost of a project.[17] Abrogation of that responsibility would have violated the NACA/NASA test and verification culture, however, and the assumptions that people from a research culture make about the discovery of truth. Skepticism is a natural part of a research culture. Good scientists do not accept the findings of their colleagues simply because their colleagues are trustworthy. Good scientists want to test and verify those findings themselves, most often by replicating the experiments that their colleagues have performed. The first NASA scientists and engineers, who adopted the NACA research culture and were relatively isolated from bureaucratic and political pressures, maintained this philosophy. *non-bureau-cratic culture*

"Our approach has always been to demand that the factories deliver a flight-worthy, flight-ready product," one NASA launch director observed. "They do all the certification and development testing and then they do a final checkout and they ship it ready to be launched, basically with the seal on it that says, this is ready to fly."

Under the terms of NASA's test culture, however, certification did not mean that testing was over. As the launch director noted, "we look at ourselves as a customer and say, okay, now before we are going to commit that to flying, we are going to connect it to everything that it has to work with [in space] that we physically can. . . . We're going to connect it here on the ground and check it out and make sure it all hums properly. And then we will service it and button it up and launch it."

"We used to take the spacecraft apart in the Mercury and Gemini program and do a thorough inspection and then reassemble it and test it." That was the culture at the Kennedy Space Center. The contractors had to "ship the hardware ready to shoot it, but be prepared that we are going to give it a thorough wring out before we commit it to flight."[18]

NASA scientists and engineers insisted on extensive retesting. "The components were retested," said one NASA scientist, "because experience had shown that such retesting was necessary to catch mistakes made by the contractors." The scientist argued that retesting saved money in the long run. "NASA's testing kept costs down," he insisted, "because it prevented failures."[19]

By the middle of the 1960s, with the Apollo program fully under way, testing procedures had become quite elaborate. "We had established formal requirements for qualification testing at different levels . . . a rigorously defined requirement for tests to be conducted," said one of the directors of the Apollo program. "I can remember, for example, a lunar module that Grumman built. By the time that the lunar module was a total flight-ready vehicle, ready to fly, there had to have been a large number—and I'm thinking probably hundreds—of formal qualification tests that would have been conducted on its component parts."

"I remember we did an awful lot of . . . qualification testing on the landing gear." The lunar module landed on four struts, with pads to keep it from sinking into the lunar surface and honeycomb compression material to soften the impact. Sensors were installed on the pads to tell the astronauts when the module was about to touch down, so that they could shut off their descent engine. "We were able to establish discipline such as 'no unexplained anomalies.' That was an important requirement in a lot of the software testing. Every novel or off-nominal result had to be fully traced to the ground." Unsatisfactory tests led to redesign work and another round of testing. "It wasn't just a matter of chance or somebody's judgment as to whether it occurred," the program director observed of the redesign work.[20] Redesign followed a rigorous set of well-defined procedures.

The same outlook dominated the ABMA rocketeers, led by the German rocket team. They were known for the rockets they developed, rather than for technical reports like those that NACA employees wrote. As rocket scientists and engineers, however, they brought to NASA the same fundamental emphasis on testing to discover how hardware worked. Testing, said one of the German engineers, makes the difference between "a successful failure and a complete failure. In one case you know what happened; in the other case you don't."

The German engineer gave an example of a successful failure that occurred during the mid-1950s, when he was still working for the ABMA. "I believe it was the first or the second Jupiter we launched. It went out of control." On account of steering movements, fuel and oxidizer began sloshing around in the tanks. Test results revealed that this sloshing had caused the loss of control. Engineers installed rings inside the tanks to break the sloshing. "One failure was sufficient. In the next designs, you see the rings inside. You know why. One failure with the Jupiter."[21] The von Braun group used a four-stage configuration of the Jupiter to launch the first U.S. earth orbiting satellite on January 31, 1958. "We made a lot of tests and, whenever something broke, we redesigned it," said another one of the German rocketeers. "We just went very much into details."[22]

The test philosophy became even more important when the von Braun team was asked to put astronauts on top of rockets. A Redstone rocket developed by the German rocket team pushed astronaut Alan Shepard into a suborbital trajectory in the first flight of an American into space. "If we want to have the . . . vehicle manned, then it has to function from the beginning," observed one of von Braun's assistants. "Now comes the detailed work. . . . When you had a test made, was it completely successful? No, here is something wrong still. Okay, we have to find that out. That has to be corrected. And that takes time." The new design would be retested. "I have now had a total of let's say

200 tests on the test stand and you know the failures very accurately. You know that you have corrected what was wrong here. Then you had a series of tests when nothing went wrong. Then you could say, okay, now I'm above that and I can trust the engine."

"This, of course, is never 100 percent." The metal out of which the engine was made could fail. "In spite of the fact that the engine had already passed two tests or three tests on the ground, you still can have a material failure." Ground tests do not simulate the flight environment precisely. "In spite of the fact that you may have checked it out completely under testing, there are always things which still slip somehow and which you find out during a flight. Not necessarily does it have to be a catastrophe. You know only: here is something wrong, which you have to check in detail. . . . None of us had 100 percent confidence from the beginning. And I always, when I saw the launch, I looked at my clock. Okay, we passed that, we passed that. The weaknesses which you knew, here, here, here. We passed that, we passed that. Now we are in good shape."[23]

With the advent of the space age, the von Braun group had to turn over much of their testing and most of the rocket fabrication work to contractors. That did not obligate the rocket team, however, to accept the contractor's work without testing it. "We trusted not even a contractor," said one. When the von Braun group received the large F-1 engines manufactured for the Saturn rocket that would propel Americans to the Moon, the leaders of the team put the engines on their own test stands at Huntsville and at their substation on the Mississippi River to make certain that the engines worked as designed.[24] The fact that the manufacturer had fired up the engines on its own test stands did not deter the group. "We put it on the test stand," this engineer said. They fired up the engine to see "what came out, how the material behaved, and if there were weak spots and so on." Testing continued at the Kennedy Space Center. "Sometimes, the companies didn't like it too much, but they agreed that if you want to have everything 100 percent reliable, then you really have to go into each and every detail."[25]

The test philosophy carried over to the satellite programs as well. In the military, said one of the NASA executives to emerge from the NACA, "if eight out of ten missiles work, you're really doing pretty good." It was less expensive for the military to buy extra missiles to make up for the ones that would fail than to eliminate the failures. "Well," the executive said, "we wanted ten out of ten satellites." During the development of the satellite, NASA would order a prototype that was used for testing but would not fly. "So I said: look, when we've finished with that prototype, let's put it on the vibration machine and shake it and rattle it and roll until it busts." The vibration tests revealed weak points in the satellite. "If you were to go back to that time," said the executive of the early days of the space program, "you would find that very, very thorough testing was the mark of the satellite business."[26]

As time-consuming and expensive as this test and verification process might seem, it was viewed as indispensable by the early leaders of the American space program. Testing achieved the status of a cultural assumption—something that agency officials assumed must be done in order to carry out their work.

In-House Technical Capability

The people who put the space program together assumed that NASA had to maintain a strong technical capability inside the agency in order to accomplish its missions. Other government agencies might allow their technical people to drift away or rely upon industries or other organizations for needed skills. NASA employees resisted that temptation. They did not want to turn NASA into a paper-pushing organization, a supervisor of other institutions from which the government purchased technical skills. They wanted to maintain those skills in-house, on the agency payroll. Such cultural beliefs, said one agency leader, were more than assumptions—they were the result of "hard-won, painful experiences with hardware that failed."[27]

The bearers of NASA's original culture sought to maintain NASA's technical capability in three fundamental ways. First, they believed that they had to perform a significant part of their space flight work in-house, that is, inside the agency using NASA employees and facilities. This belief conflicted sharply with the reality that most of NASA's space-craft and satellite work, particularly the fabrication of hardware, had to be done by contractors working outside the agency. Seeking an appropriate balance between in-house capability and contractor responsibility was one of the most taxing cultural problems that NASA officials faced.

Second, NASA officials made a tacit promise to their employees that they would have the opportunity to do hands-on work. That is, NASA employees would have the opportunity to build satellites, launch spacecraft, and control space flight operations. Hands-on work sharpened technical skills and in the process expanded the capability of NASA employees to monitor the technical work of contractors.

Third, NASA officials sought to maintain technical capability by recruiting people with exceptionally good technical skills. This could only be done if the agency provided its employees with the opportunity to do hands-on work using their own facilities. Exceptional people, the NASA leadership believed, would not join the space program unless it offered them the chance to actually work with the machinery.

The underlying desire for in-house capability was inherited from the NACA and the ABMA. Both possessed a tremendous amount of in-house technical capability. Because the scope of their projects was small, the NACA research laboratories were able to do most of their work in-house. This allowed the laboratories to develop over time a strong corps of professional and technical employees. "We had our own engineering shop," said one Langley engineer. "We had our own sheet metal machine shop, plastics, wood shops. We could build anything there, provided it wasn't too big. We had a great model shop." The Langley Research Center was famous for its model-building shops. "We did it all ourselves."[28]

NACA employees who went on to become NASA executives insisted that their in-house work had trained them for the difficult task of space exploration that followed. Training its employees through in-house work "was a very, very significant aspect of the NACA," said one of the Langley employees. "I had to do every part of the job," he said, recalling how the NACA's in-house work had prepared him for the space program. As a NACA employee, he had been assigned to manage a research project on aeronautical stability and control:

all in house

I had to write its requirements, I had to provide its instrumentation. If anything had to be built, I had to design it and I had to take it through the shops and assure that it was built properly. . . . I had to see it was properly installed in the airplane. I had to deal with the safety of the airplane. I had to deal with the mechanic of the airplane, the crew chief. I had to write the flight test. I had to assure that it was not going to destroy the vehicle I was playing with.

I had to interface with the pilots. I had to tell them what I wanted done. I had to reduce the data by hand. If there was any calculation to be done, I had to do it myself. If it was a hand computer, I did that; and eventually an electronic computer, I had to do that myself. Then I had to write the report. And then I had to defend it in a court of review.

If there was any interface with the outside world in the way that airplane was built, or the way it flew, or the improvements that needed to be made, I was the interface with the outside world, with the people that built that airplane.

There is no other place in history where that kind of training could be achieved. As a result, when I got to NASA, when we started the

Space Task Group and I became a manager of people and a manager of contracts and a doer as opposed to a thinker, there was nobody in the United States that could snow me. Nobody. I don't care how good he was or how much smarter he was than I was, the training I had at the NACA prepared me for dealing with that kind of situation. And believe me they all try to snow you.[29]

The ABMA was organized in a similar fashion. It was, said one of the leaders of the Apollo program, "a powerful technical and engineering organization. . . . Their mind set was largely, do it yourself and build it in-house. To do major contracting . . . was foreign to their experience."[30] On the grounds of the Huntsville facility, ABMA employees conducted their own research and development programs. They created their own fabrication shops and hired their own craft employees. They built their own prototypes. They even built their own rockets. ABMA employees built the first Redstone rockets, the vehicle used to fire the first American astronaut on a suborbital trajectory. The first seventeen rockets were fabricated and assembled in government facilities, after which the Chrysler Corporation began manufacturing the rocket at its own plant.[31] Turning a well-developed project over to a contractor was not contrary to ABMA traditions, provided that ABMA employees maintained tight control over the work. When a contractor took over production of a rocket, or produced a component like a rocket engine, the ABMA rocket team closely supervised the contractor's work.

This was a well-established tradition, both within the U.S. Army, for which the German rocket team worked, and back in Germany. Within the U.S. military, the ABMA came to epitomize what was sometimes known as the arsenal approach to weapons development. The U.S. Army had been manufacturing and storing munitions at government arsenals since the republic began. The Redstone Arsenal, to which the von Braun team came, was established in 1941 as an army ordnance plant for the manufacture of chemical compounds, pyrotechnic devices, and small rockets. At Peenemünde, the German rocket team had worked as part of a complex web of research laboratories, assembly facilities, and privately owned manufacturing firms. The research laboratories at Peenemünde were part of the German military, while the manufacturing plants worked under tight government supervision.[32]

When the von Braun group began work on the Saturn I rocket—the predecessor of the Saturn V that would lift Americans toward the Moon—they began with considerable in-house capability. The von

Braun team built the first eight Saturn I first stages at the Huntsville facility. At their own laboratories and test facilities, they made improvements in hardware and checked over the work of contractors as the latter became involved. In late 1961, the Huntsville group—by then the NASA Marshall Space Flight Center—finished the basic design work on the Saturn V first stage and awarded an industry contract for its construction. The contract went to the Boeing Company. The government rocketeers supervised the design of the vehicle from the Huntsville center and insisted that Boeing do much of the manufacturing work at an annex of the Marshall Center known as the Michoud Assembly Facility, in Louisiana, under the direction of center employees.[33]

To a group of congressmen in 1960, von Braun explained the management philosophy that underlay his group's commitment to in-house capability. "There is simply no industrial corporation in this country that can competently tackle all problems involved in a vehicle like the Saturn," von Braun argued. The Saturn rocket involved state-of-the-art engine technology, inertial guidance and rocket control, automated computer systems, the handling of liquid hydrogen, and explosive devices for stage separation. "So Uncle Sam has to go to a great many contractors, if he wants to utilize his national resources and have American industry make an optimum contribution. But this puts the burden of coordinating such a program on the back of government agencies."

"Experience has shown time and again that in government agencies you cannot build up and retain competency over any length of time unless you give government personnel the possibility to keep in intimate touch with the hardware and its problems." This was the reason, von Braun explained, for maintaining an in-house rocket development capability at Huntsville. "If we would convert Huntsville to an all-out contracting operation without any continued in-house work, our best people would soon run away and say: 'Here I get rusty. I go to where the contracts go, because that is where the interesting work is done.' Soon we would have no capability left to coordinate the overall Saturn effort."[34]

NASA officials who favored a strong in-house capability offered other arguments supporting their point of view. In-house work helped to create institutional memory. "You've got to keep the experts around who know that system," said one of NASA's top engineers. Industrial firms scattered their workers about when a project was finished. "Those people disappear and it is a lot harder to get them back. When you have got them as in-house civil service engineers, you've got more flexibility in pulling them off the job they are on and saying hey, we need your help for four weeks over on this satellite that you helped design ten years ago."

"I will give you a very concrete example," he explained. "The tape recorders in the [Hubble] Space Telescope are a very old vintage." They were manufactured by IBM around 1980. The telescope was launched in 1990. "If that thing is still operating fifteen years from now, we are flying at best a twenty-five-year-old technology." If something goes wrong, NASA will need to assemble a team of people who understand how the recorders work. "You won't get that by pulling the drawings out and looking at them." NASA will need to talk to the engineers. "Where are the people who designed that stuff twenty-five years ago?"[35]

In-house work also promoted flexibility. One of the German rocketeers explained how this was so. The rocket team rarely came up with a design that worked from the start. "You have a failure on the test stand somewhere. That has to be corrected fast." With an in-house shop, the rocket engineers could go directly to the shop employees and tell them to start work. They would not have to make drawings. They would only have to explain the problem. "You have an immediate correction. That's a matter of hours, maybe a matter of one, two days. If you have [to go to] a company, you have to make a drawing, you have to make an estimate and it takes—all of a sudden—at least two weeks, which we did not have." With in-house capability, "we always could correct immediately." The ability of the German rocket team to develop reliable launch vehicles required such capabilities, he insisted. "Flexibility is very, very important in every new program which you have."[36]

In spite of the underlying tradition of in-house work, it was clear from the beginning that the new space agency would rely extensively upon contactors. T. Keith Glennan, the first NASA administrator, wanted to contract out the bulk of NASA's work.[37] The second administrator, James Webb, wanted to use contracts to build up the aerospace industry and university science and engineering programs. NASA officials recognized that the agency did not possess sufficient facilities for manufacturing the spacecraft and rockets they needed. Furthermore, not everyone associated with NASA believed that effective program management required in-house capability. Pockets of resistance to the in-house tradition emerged. The U.S. Air Force had developed a program management system that relied more upon contractors, and as Air Force officers joined NASA to help launch the Apollo program, they promoted their system of contractor responsibility. Aerospace executives in private industry, confident of the technical capabilities of their own companies, understandably wanted to see a greater role for contractors.

The Air Force system contrasted sharply with what many viewed as the arsenal system for hardware development. In the development of its intercontinental ballistic missile program in the 1950s, the Air Force relied upon contractors not only for the design and manufacture of missiles, but also to perform a great many coordination and management functions. "The Air Force," wrote one commentator, "turned to private contractors because it had neither the depth of competence found in Army laboratories nor the time to recruit engineers."[38] The Air Force system allowed for a rapid buildup of technical capability, albeit in industry, with a minimal government staff. In some circles, the Army arsenal system was viewed as outmoded, a relic of nineteenth-century munitions policies. These people viewed NASA's tradition of in-house work and substantial contractor oversight as a temporary necessity that could be set aside as soon as industries and universities built up their space expertise.[39]

Bearers of the original culture searched for a proper balance between old traditions and the new reality. "It was quite a traumatic thing," said one of NASA's new center directors. "NASA was going to do a lot of contracting out. They were going to pay industries to do things, and that was something that the NACA had never done. Many people in the NACA thought that this was going to be the downfall of the new [agency]. On the other hand, there was no way you could run a space program and put satellites in orbit without bringing industry in and paying them to build those things, because certainly you weren't going to build all those things in government laboratories."[40]

Von Braun developed a general rule: 10 percent of the money spent on rocket development should be spent in-house. "This is the only way," he announced, "to retain professional respect on the part of our contractors."[41] The 10 percent, while a seemingly small fraction, would be sliced from a very large pie. NASA spent nearly $10 billion to develop and build the Saturn rocket. The spacecraft that took the astronauts from Earth to the Moon and back cost $5.5 billion through 1967.[42]

Officials at the Goddard Space Flight Center, who also relied upon contractors for the bulk of their satellite work, developed a similar rule. According to a leading Goddard executive, the center tried to have at least one or two small satellites under development in-house at all times. They also sought to perform a major share of the work on one of the large satellites in-house.[43]

"The basic policy was that NASA would do enough missions in-house so that it always would have the technical capability in-house to judge a contractor accurately and to bail a contractor out if a contractor got into trouble," said one of NASA's leading scientists from the Apollo era.[44] An early consulting report made a similar observation. "Three general criteria—never officially promulgated—have guided the development of the in-house capabilities of NASA's Space Flight Centers." Each center would develop enough in-house capability so that it could do the basic conceptual work on new projects, develop technical specifications for private contractors and supervise their work, and ensure the technical excellence of its staffs. Flight centers would rely upon NASA laboratories for their advanced research.[45]

Such policies, written and unwritten, never resolved the underlying conflict between the tradition of in-house work and the reality of big-time contracting. Some officials, particularly at the old NACA laboratories now become NASA field centers, resisted the push to contract out. "One of the rules I had," said one early research center director, "I would not contract at all on any research program. I just didn't do it. What I did was, I did a little contracting of things like window washing and repairs of buildings, building maintenance and stuff like that. Guards. But I would not contract out any research activity or any technical activities. The reason to me was very clear. I was throwing away my future. The way you keep these laboratories alive is to grow your own people in them in the right environment. . . . As soon as you stop doing that, you become a bookkeeping organization."[46]

In late 1961, NASA announced the award of the contract to build the second stage of its giant Saturn V rocket. It was a tricky project. Unlike the first stage of the Saturn V, which used a conventional kerosene-type fuel, the S-II burned a much more powerful mixture of liquid hydrogen and liquid oxygen, a state-of-the-art technology. Unlike the Boeing contract given out for the Saturn first stage, the contract for the S-II allowed the contractor to assemble the rocket in Seal Beach, California, far from Marshall Center in-house facilities.

The relationship was a troubled one. One of the first models of the S-II to come out of the assembly plant ruptured on the test stand. So did a second test stage. Delays on the S-II program threatened to push the lunar landing past the much emphasized end-of-the-decade goal. "It is just unbearable," said one of the German rocketeers from the old ABMA, "to deal with a non-performing contractor who has the government tightly over a barrel when it comes to a multibillion dollar venture of such national importance."[47]

Headquarters officials pressured the von Braun team to give the contractors more latitude. The German rocket team complained and dug in its heels. One of the top officials from the headquarters team explained that "[NASA Administrator James] Webb and other people raised hell with them because every single engine that [the von Braun team] got from North American they took apart piece by piece by piece and then put it back together." (North American Aviation had received the contract to build the S-II.) "Webb said, you know, we don't do that. We've got to trust American industry. He was a little cocky about things like that." The issue reached a head in a meeting called to resolve the dispute. A member of the von Braun team produced a rag. "He says, this is what we find in here, Mr. Webb, this is what we find in this stuff."[48]

Like other oral traditions, the event was embellished to become a story that emphasized the moral it contained. Different people told different versions. The story provided a defense for von Braun's desire to closely inspect the work of his contractors. In one celebrated instance, contract workers at what became the Kennedy Space Center went out on strike because the von Braun team would not leave them alone. The workers were accustomed to Air Force practices, which involved little direct supervision. They complained that von Braun's minions had installed cables and consoles at the launch site. The Secretary of Labor had to appoint a fact-finding committee to entice the union members back to work. The committee never resolved the basic differences between the von Braun team and the contract workers.[49]

Much of the insistence on in-house work and contractor oversight was based on the pride that the NACA and ABMA employees had in their own capabilities. It might seem like technical arrogance to someone outside the agency, but the NACA and ABMA leaders really believed in their technical abilities. The attitude of many NASA employees from the NACA and the ABMA toward their contractors was like that of teachers toward their pupils. "Everything was taken apart and reassembled," said one of the German rocketeers. "First to see their workmanship and, second, to show [the contractors] how they should be better. And our best technicians did this kind of work."

"I made as a first condition to the company that they do not make modifications on our design." The rocket team insisted that the contractors first learn why the team had designed it the way that they had. "Then we would be open minded." One company got a contract to build a control motor for the Redstone rocket. It was "a unique motor, refined in every respect." When company officials produced their plans for the motor, "it was so changed that it was impossible to accept it. That was the typical way some companies tried to work with us. We recognized this very quickly and that is why we demanded at that time that they learn from us first." After they showed themselves to be good pupils, then the contractors could make suggestions. There was an added reason, based on in-house capability, why von Braun's lieutenants wanted the contractors to follow their design. "Since we had built it ourselves," he said, "we knew it could be done."[50]

Hands-On Experience

Another way that NASA officials maintained the agency's technical capability was by offering their employees the opportunity to do hands-on work. NASA employees sat at consoles at Mission Control. NASA astronauts flew the spacecraft. NASA engineers worked out the design problems on new spacecraft and satellites. NASA scientists solved basic flight problems at the research centers. In spite of the fact that contractors played a major role in the civilian space program, NASA employees fought to keep their hands on the hardware. The desire to do hands-on work was an essential element in the overall culture. NASA employees viewed hands-on work as a principal means for strengthening their basic technical skills. They viewed it as a training ground for people who would have to supervise contractors. They saw it as one of the primary means for attracting exceptional people to join the work force, by making sure that their work was exciting.

"We never want to become an agency that is just a contract monitor," said one center director. "That ain't much fun. The real joy of this business is actually doing it. . . . The whole place thrives on . . . engineers and scientists actually doing, participating in the program, not pushing paper around watching somebody else do it—somebody else being anybody outside the civil service. That's a pretty ingrained piece of the culture."[51]

To most people involved with NASA, hands-on work was synonymous with in-house work. It was the opportunity for NASA employees to work a project from start to finish without having to turn it over to contractors. NASA employees believed that this sort of hands-on work gave them the ability to recognize incipient technical problems when someone else—such a contractor—started to make them.[52] Some believed that hands-on work prepared NASA employees for broader managerial responsibilities. Within the NASA culture, these were strongly held beliefs.

The hands-on culture went beyond this. It went beyond the opportunity for NASA scientists and engineers to work alone on in-house projects. It also characterized the nature of space flight operations and the relationship of NASA civil service employees to their contractors. An official from the Kennedy Space Center, which relied extensively upon contractors to prepare rockets and spacecraft for launch, offered this definition. "The thing that makes NASA different," he said, "is that government employees of NASA get much more involved in the intricacies of every project or every task than Air Force or other civil servants do with their contractors."[53]

As with many motivating beliefs from the original culture, faith in the value of a hands-on organization grew out of the experience of the NACA/ABMA days. "The thing about the NACA was that there was a great number of very small projects current at any one time," said a leading NASA engineer. "Consequently, there were a large number of what you might call project engineers, each completely in charge of a separate project. You would be in charge of getting the job done through the amount of money that had been allocated to it. Make sure that the model was strong enough to hold together. That it was stable and wouldn't get over-heated and burn up. You name it. You had to make it work."[54]

One NASA employee wistfully recalled how the process worked at an old NACA research center. "It was just about what I thought it would be. Very research oriented, a tremendous amount of flexibility for even the youngest, greenest researcher. . . . It was look, here's a job, here's a task and here is a research project. You decide what you want to do and how you are going to do it." This employee was told to construct a laboratory project that would simulate the impact of high-speed objects, such as meteoroids, on objects in space. "I would do the design. Okay, that's great. I would then make out what we called the work order, which would say please construct this piece, made of such and such a steel and all that good stuff. I would take that work order and my design and I would go over to the machine shop, which was just across the street." All of the NACA research centers employed machinists and electricians. "They were a critical part of the team. . . . Typically, a team was the machinist and the electronic technician and me. . . . It was real time, none of this, you know, put out a procurement to have this thing built on contract, which comes back in three months."[55]

Another employee who eventually became a NASA center director remembered the work he did as a NACA researcher at the Langley Aeronautical Laboratory during the 1930s. "It was the work I did on downwash," he recalled. (Downwash is the downward deflection of air created by an airplane wing.) "We put an actual airplane, a number of airplanes, in the [wind] tunnel. Each one of course had its own wing shape and wing size and we had a big overhead crane that rolled this way and that." A technician controlled the airplane using chains. "You would just sit there reading the manometer," (an instrument used to measure the pressure of gases or fluids). "In those days we didn't have the automatic recording manometers and those sort of things. We read the manometers." They ran the tests at night because the electric power that drove the wind tunnel was cheaper at night. "Talk about 'hands on,' that was very much 'hands on.' . . . We worked right with the equipment. No, there were no contractors or anybody around the place at all. It wouldn't be tolerated."[56]

The tradition of hands-on work continued after NASA was formed, even at the centers that contracted out the bulk of their work. In Project Mercury, which propelled the first Americans into space, NASA engineers designed the single-seat space capsule. NASA engineers conducted the first airdrop studies to determine the aerodynamic characteristics of the capsule returning to earth. NASA engineers conducted the first tests of the escape rocket that could jerk a space capsule away from the launch pad if an accident occurred. Engineers at the Langley center and at the Ames Research Center in California put models of the blunt-body Mercury capsule into their wind tunnels to test drag, lift, pitch, vibration, and flutter. Langley engineers participated in the first parachute tests, examining how the space capsule would land. Tests like these kept NASA employees directly involved in the mechanics of space flight.[57]

At the newly created Goddard Space Flight Center in Beltsville, Maryland, NASA scientists and engineers established a tradition of building their own satellites. Of the twelve Explorer series satellites managed through the Goddard Center between 1969 and 1975, for example, fully half were built by Goddard Center employees.[58] A NASA scientist who became a Goddard Center director attributed the success of their satellite programs to the tradition of hands-on work.

They had a lot of hands-on people right there that all worked together: the scientists, the engineers, the operations people. . . . You could go out there and build a satellite, launch it, operate it, do everything that needed to be done. . . . To have a community of people that understand things to that degree—end to end—puts you in a very good position to oversee contractors. Ninety percent of Goddard's money really goes through the center and out to contractors, but they are always doing an in-house project . . . building some instruments in house and doing those kinds of things that keep you smart.

They have got machine shops. . . . They can build anything that you need to build to put on a spacecraft. And there are technicians . . . in the labs, building instruments and experiments. They have everything that private industry would have. . . . They would buy parts from various people when they needed to do that, but they could build all the other things they needed there. Integrate it, test it, check it out, and launch it. . . . That brings an interesting culture.[59]

The faith in hands-on work persisted into the second and third decades of space flight. One NASA engineer, who later went on to direct the Moon/Mars initiative, got his start with NASA in the 1970s as a subsystem manager studying the problems involved in separating solid rocket boosters (SRBs) from the space shuttle. "I didn't specifically have any hardware that was directly my responsibility," he explained. He was responsible for integrating the way that the shuttle orbiter, the external fuel tank, and the boosters behaved as the boosters moved away two minutes into the launch. "The proximity aerodynamics turned out to be very complicated." He and his assistants studied the aerodynamics of SRB separation in a large wind tunnel.

"The tests were fascinating." Solid rocket booster separation takes place at the upper edge of the stratosphere, twenty-eight miles above the earth, at a point where the shuttle has accelerated to a speed of 3,100 miles per hour. Enough atmosphere remains at that altitude to create some fairly complex aerodynamics. "As the boosters move off," the engineer explained, "their aerodynamics influences the aerodynamics of the other vehicles. . . . It's additionally complicated by the fact that there are solid rocket motors on the boosters that blow them away from the orbiter." When those rockets start to fire, they create plumes that alter the flow of air around the orbiter. "It's a very complicated test. It involves the relative motion of the models, it involves simulating these big plumes—so we spent a lot of time developing techniques to do that and then getting the data." This engineer also led the team that developed the computer program on the orbiter that controlled the whole sequence.

"It was for the most part hands-on work," he explained. Rockwell International had the contract to build the orbiter, but NASA took charge of the separation aerodynamics. The people who worked for the contractor "ended up essentially being an extension of my staff," a team of people working together on the problem. "It was a good way to cut your teeth," he recalled. He was twenty-six years old at the time.[60]

NASA officials argued that the extensive hands-on experience employees received in the predecessor organizations gave the new space program the quick start it needed to catch up with the Soviet Union. "Examine other groups that have tried to start up from scratch, in the government or any place else, and you will find that they have trouble," said one early center director who had worked for the NACA. "We were able to start off with kind of a running start because of the talent we had amassed. It's unusual to be able to pull together groups of this extreme talent."[61] When asked why the rocket team that formed the nucleus of NASA's Marshall Space Flight Center pulled ahead of the Vanguard group with whom they competed to launch the first earth satellite, one of the German engineers replied: "It's very simple. It's experience." Hands-on activity gave them more experience. "All the mistakes they made we had made before."[62]

To understand how hands-on work influenced projects where contractors were involved, it is necessary to understand how NASA organizes its field centers. No center is typical, but the Johnson Space Center illustrates the basic philosophy. The directors of the Johnson Center deploy their engineers and scientists into three principal areas: technical work, project management, and operations. The hands-on work conducted in the technical area allows project managers to maintain close control over the details of contracted work. It allows the project managers to keep their hands on the work of contractors.

When the Space Task Group that managed the Mercury space flight program moved from the Langley Research Center to Houston, it brought with it an important organizational principle from the old aeronautical laboratory. More than one thousand employees from the Langley Research Center moved to Houston to form the nucleus of what they called the Manned Spacecraft Center.[63] The first director of the Manned Spacecraft Center, Robert Gilruth, who had headed the Space Task Group at Langley, set up a large engineering directorate at Houston. (Whereas a directorate generally refers to a small board, NASA engineers used the term to describe a large department.) The engineering directorate at Houston replicated the old laboratory system at Langley, which contained research divisions in areas like stability, structures, loads, and the compression of air. It contained divisions that investigated problems like guidance, power, structures, and environmental control. The directorate was headed by Max Faget, a NASA engineer widely regarded as the key person responsible for the design of America's first manned spacecraft.

It was clear to all involved that the Houston center would rely extensively upon contractors for its space flight programs, especially for the fabrication of spacecraft. To coordinate the development of the Mercury, Gemini, and Apollo spacecraft and oversee contractors, the center directors appointed project managers.[64] (NASA later changed the nomenclature so that very large systems like the Apollo spacecraft were directed by program managers.) The project management technique was invented before NASA began, to coordinate high technology activities such as the development of military ballistic missile systems. The approach operated on the principle that responsibility for the timely completion of a project could be concentrated in a single management team that would coordinate all parties making contributions to the undertaking, including those who did not officially report to the project manager.[65]

Many of the engineers and scientists employed at the Johnson Center during the 1960s worked within the project or program offices. An equally large number, however, worked within the engineering directorate.[66] The project managers asked people from the engineering directorates to work with them on the mission. The project management approach sought to create exactly this sort of situation—people from functional divisions working hard on behalf of mission objectives, even though they reported to separate bosses.

Within NASA, the engineers and scientists in the technical branches made a special contribution to the project management offices. They provided the technical know-how that allowed the project managers to monitor the work of the contractors in depth. To provide such technical advice, the engineers and scientists had to acquire hands-on experience. The director of the Apollo Spacecraft Program Office at Houston explained how the system worked. His office was responsible "for the overall management of the program." It took care of the business and financial aspects of contracting and it looked after "the overall technical aspects of the program . . . all the different technical disciplines that are required at the subsystem level in order to do the actual work."

Rather than bring the technical specialists directly into his office, "what we do is draw very heavily on the rest of the organization for specialty activities. . . . With Max Faget's directorate for engineering and development we have all of what we call the subsystem engineers." They were responsible for subsystems such as environmental control, the fuel cells, and spacecraft propulsion. "They reside in the engineering and development directorate for several reasons, not the least of which is they are with their fellow technical specialists." Although they reported to Max Faget, they responded to the needs of the Apollo program management team.[67] For really complicated activities, NASA program managers would also call upon specialists from research centers like Langley, Lewis, and Ames.

A similar system evolved with the rocket team at the Marshall Space Flight Center. An industrial operations branch housed project managers for launch vehicles like the Saturn V. A research and development division housed the technical labs. "In a program as novel as ours, it must be expected that there will be setbacks and problems in development," the center director, Wernher von Braun, once explained. "You cannot simply write a final contract on a stage of the Saturn V and let the contractors go." The contractors ran into difficulties; they came up with new ideas. Sometimes NASA engineers wanted to make changes.

"Administration of these kinds of R&D contracts means riding herd on a lot of technical, managerial and financial details," von Braun continued. "But no project director . . . can possibly be an expert in all the technical and scientific disciplines involved. The real mechanism that makes Marshall 'tick' is therefore a continuous crossfeed between the right and left side of the house." On one side, officials managed the projects. On the other side, "we have the knowledge in depth." The project director could call in the experts "to help him do the job, pass judgment, and monitor any critical phase of the contractor's operation."[68]

NASA officials from the original culture believed they needed to provide their engineers and scientists with hands-on experience in order to maintain the technical side of the house. It was the only way to keep them technically sharp. By keeping their own engineers and scientists sharp, they could penetrate the work of the contractors.[69] "It puts the government in an over-the-shoulder, detailed monitoring role," observed one official from the Houston center. NASA civil servants, he said, go "very deep into everything [the contractor] does, from engineering, to his production planning, to his reliability, to his test program, to the acceptance tests."[70]

The loss of that sort of hands-on work made the bearers of the early culture anxious, if for no other reason than it diminished their ability to monitor contracts. The technical specialists contracted their work out too, just like the project managers, a development of much concern to the first generation of NASA managers. "You know," said one director of the Johnson Center, "if a guy is going to tell somebody how to build an APU, an auxiliary power unit that is going in a spacecraft, you can't do that by learning on paper. You need an APU of your own. You need to build one. You need to work on one. You need to take it apart. You need to test it. You need to understand its idiosyncrasies and its faults and its virtues. And unless you have your hands on the hardware, that doesn't happen."[71]

Hands-on work also played a key role in the third major area into which the managers of the Houston center deployed their staff: space flight operations. The people who directed expeditions into space from Mission Control developed a subculture all of their own. They had their own language. (A Saturn rocket on its way to orbit, for example, "went up the hill.") They had their own rules (describing, for example, the circumstances under which a flight crew should abort a mission and return to earth). They had their own legends and their own heroes.[72]

One rule was so deeply embedded in the subculture of flight operations that no one bothered to write it down.[73] NASA ran Mission Control with NASA employees. The people who planned space flight operations at Houston were NASA employees. The people who sat in the "Trench" were NASA employees. (The front row of consoles at Mission Control was known as the Trench, particularly the consoles of the flight controllers who oversaw the navigation and the firing of rockets guiding the astronauts through space.) The job of keeping the astronauts alive while they were in space was a NASA responsibility. Contractors might work in the back room. They certainly provided technical advice. But the responsibility for the mission was something that NASA under the terms of its original culture did not dare to share.

During the 1960s, a similar tradition prevailed at the Kennedy Space Center. NASA employees controlled the final preparations for human expeditions into space. Contractors, to be sure, checked out and assembled rockets. A NASA official, however, directed the launch. NASA employees controlled and coordinated the contractors. NASA employees sat in the front row of consoles at Launch Control.

The hands-on tradition was deeply ingrained in the NACA and the ABMA and was one of the distinguishing features of NASA during the early era of space flight. Other government agencies might allow their employees to become paper pushers and monitors of the people who did the actual work. NASA officials, under the terms of the original culture, were determined to keep their own employees close to the work of space exploration and flight research—even when contractors were involved.

Exceptional People

People of average mental and physical capabilities staff most large organizations. Recruiting practices bring to those organizations a random distribution of employees that approximates the familiar bell-shaped curve of the work force at large—some exceptional people, some below average, but most in the middle. This is especially true in government. Most government executives do not possess the tools necessary to attract an exceptional work force—the ability to dispense challenging assignments, high pay, or control over one's work. Public executives face grave difficulties in attempting to maintain for any substantial period of time a sizable work force that is significantly better than the population from which they must recruit.

An important part of the NASA ethos denied this reality. The performance of the American space program, according to the people who ran it, was based on the view that NASA had been able to beat the odds and assemble a work force made up of highly talented engineers and scientists. This was one of the most persistent beliefs in the early NASA culture and remained so as the agency matured. From the beginning, NASA managers maintained a policy of recruiting the "very best" people they could find for key positions.[74] Results from a survey conducted as part of this study in 1988 revealed a continuing faith in NASA's ability to recruit outstanding people, even after the traumas of budget cutbacks.[75] (See the appendix for survey results on recruiting exceptional people.) A 1990 commission report on the future of the U.S. space program identified exceptional people as one of the two "most fundamental ingredient(s) of a successful space program."[76] Within the NASA culture, exceptional performance and the maintenance of technical capability depended for the first thirty years upon the agency's ability to recruit outstanding people.

As with many cultural norms, this belief took the form of an assumption. NASA officials assumed that the success of the civilian space program could be traced to their ability to recruit exceptional people. Agency officials, of course, saw this as more than an assumption. They believed it was true. They believed they actually did recruit extraordinary people. Accordingly, they placed demands on their work force that they could not have made if the work force was merely adequate. From outstanding people they demanded outstanding performance. In particular, NASA management demanded that their employees work harder than average civil servants and pay more attention to detail than was expected in a typical government organization. NASA employees, believing themselves to be part of an exceptional group, responded positively. In this way, management's faith in the quality of its people was affirmed. Higher expectations, as a consequence, helped NASA officials overcome many of the obstacles that deter high performance in government agencies.

Faith in the exceptional quality of the work force existed at the beginning of the space program. NASA officials believed that they had started with a talented group of people recruited by the predecessor organizations. "That was the secret of NASA's success," said one agency executive. The predecessor organizations had already recruited outstanding people, starting with the NACA. "The NACA basically hired the cream of the crop from colleges—the intellectuals—because they had a reputation for that. . . . Then they gave you great responsibility almost immediately. A very young man in the NACA could get access to facilities and make decisions on carrying out his own program far beyond the responsibilities you get initially in industry."

Government agencies often thrust large responsibilities on young people. Starting with what they believed to be exceptional people, NACA managers enlarged the skills of those people by giving them challenging job responsibilities. The NACA people, as a result, "had a lot of confidence that there was no technical problem in flight that we couldn't understand. And we didn't do anything without understanding it." The ABMA and Jet Propulsion Laboratory, also brought in to help form NASA, "had the same type of culture," the executive added.[77]

Another NASA executive who came out of the NACA made the same argument. He had helped to organize the Ames Research Center in California in 1940. "We were in the recruitment phase," he recalled. He had reviewed college transcripts from engineers and scientists being considered for positions at what was then called the Ames Aeronautical Laboratory. "We could pick the cream of the crop from Cal [the University of California], from Stanford, and Washington [the University of Washington]."

In 1959, the same official helped to organize NASA's Goddard Space Flight Center. Goddard was created around a nucleus of two hundred scientists and engineers from the Naval Research Laboratory, most of whom had been working on the Vanguard project. "There is no other place in the United States where we could have assembled a staff such as we did, in part because of the surrounding government institutions. . . . I didn't realize it when I started out, but found out very quickly that I had got about 50 percent of the people in the United States that knew anything about satellites." The Vanguard group was responsible for developing both a rocket and a series of small satellites that could study Earth and the space around it. Of the six Vanguard missions attempted, four failed, including the first effort to launch an American satellite. "Yeah, you can say, Vanguard, it's not famous for what it accomplished. But the fact of the matter is these guys had been through the mill. They really knew what was to be done."[78]

Why would exceptional people come to work for NASA? To start, NASA officials attempted to pay salaries competitive with those paid in the aerospace industry. They were not always successful, but they tried. The National Aeronautics and Space Act of 1958 authorized NASA Administrator Keith Glennan to establish 260 positions for which NASA could pay salaries, according to Glennan, as nearly as possible competitive with those being paid "in the best modern research and development organizations in industry." These were what the government called "excepted" positions—taken out of the usual constraints on hiring and salaries imposed by the central personnel agency. Congress authorized the positions, Glennan said, "solely to attract and retain the specially qualified scientific, engineering and administrative personnel necessary to maintain this nation's leadership in aeronautical and space activities."[79] In all, NASA officials were allowed to fill more than 700 positions not constrained by federal pay regulations during the formative years.[80] The desire to avoid civil service constraints in areas like salary policy was a continuing theme within NASA. The theme was long on desire and short on reality, however, as government pay typically lagged behind industry standards.[81]

Second, NASA officials tried to foster a relatively free transfer of personnel between the civilian space agency and the other institutions in which the nation's scientific and engineering elite resided. During the formative years of the space program, it was relatively easy for people to move into NASA from American universities or the aerospace industry and back again. NASA did not fit the popular image of a government agency staffed by bureaucrats with a lifetime sinecure. The American space program was run by a citizen army, a collection of the best talent to be found in the nation at large. Of the people who were hired during the Apollo era and stayed, nearly half had already started careers with industry before joining NASA (see the appendix: survey results on working for industry). Many others served at NASA for a few

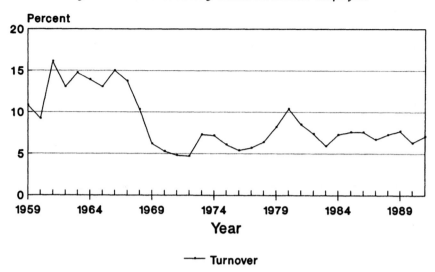

Figure 2   Turnover among NASA Permanent Employees

—— Turnover

Sources: Jane Van Nimmen and Leonard C. Bruno, *NASA Historical Data Book*, vol. 1, *NASA Resources, 1958–1968*, SP-4012 (Washington: NASA, 1988); and NASA Office of Human Resources and Education, "The Civil Service Work Force," annual (Washington).
Note: Turnover is measured by the number of voluntary separations as a percentage of start of fiscal year employment. Reductions in force (which took place from 1971 through 1975) are not included, nor is the transfer of the Electronics Research Center to the Department of Transportation (1971).

years and then moved on to other careers. In 1964, when NASA officials hired 6,200 new permanent employees, 3,900 old employees departed for other pursuits. High turnover rates continued through NASA's period of expansion. NASA had an active revolving door.

Competitive salaries and the relative ease of moving in and out did not by themselves attract outstanding people to the civilian space program. They merely removed obstacles that might otherwise have prevented it. What motivated people to join NASA was the challenge of the work.[82] "NASA was able to attract a large number of very experienced, competent, capable people in all disciplines," explained one program director who joined NASA during the 1960s, "because of the attractiveness of the program, the challenge [and] personal reward involved." It attracted people from industry, from universities, and from other parts of the government.[83]

"The government needed good people," said one NASA official who had worked for private industry on the Air Force missile program. "You couldn't just stay in industry." He had said to himself, "If I ever get asked to go into the government, I will do it. . . . I felt like that was the right thing to do." In spite of NASA's desire to pay competitive salaries, the official took a 50 percent cut in pay when he joined NASA. "There was a certain amount of 'do what you can for your country,'" he said, referring to John F. Kennedy's challenge to Americans in his 1961 inaugural address. "It was an interesting program and a chance to contribute," said the official.[84] After five years with NASA, he returned to his industrial career.

"It was the mystery . . . the thrill and the audaciousness of the task," said another NASA executive recruited from one of the industrial firms that supported the Air Force missile program. "NASA had been given just an incredible responsibility and freedom to carry that responsibility out," he added. He was awed by the people who worked for NASA. "It was difficult not to get a feeling of just great respect for the intellect that NASA represented or the imagination that the people had who were working for it." The people were willing to make decisions and move forward. "I don't know if this is a good comparison," he said, but it was almost like a war "in the challenge and responsibility that it put on individuals."[85] The space race during the first decade of exploration was very much a part of the cold war.

Much of the fascination arose from the nature of the machinery itself. For NACA employees, it began with the romance of the airplane. The Wright brothers made the first powered flight in December 1903. By 1920, NACA engineers were conducting aeronautical research at Langley Field. "The people that worked in aviation at that time were really in love with the idea of developing the airplane," said one NASA center director who started out at the NACA. "They needed to get a salary, but they would have worked for nothing if they could have had some other way of eating and sleeping."[86] NACA pay did not match what an engineer could receive from industry, but the challenge of the work in many cases made up for that.[87]

Many NASA employees, especially after the decision to go to the Moon, were attracted by the lure of space travel. "It was sort of Buck Rogers," said another official who jumped from industry into NASA, "but it was for real. This stuff is real. You're not going to get any excitement like that for the rest of your life. . . . It was exciting as hell."[88]

ABMA employees, before they became part of NASA in 1960, had been confined for most of their careers to the development of military missiles that could propel payloads from one point on the ground to another. Not until 1957 did they receive official authorization to explore space. Some of the members of the original German rocket team drifted away after they reached America, taking jobs with private industry, where they could get better pay.[89] Wernher von Braun struggled to keep alive their long-term vision, which was the development of rockets that could reach into space. "I think that it was in 1955," said one of the German engineers, "that Wernher von Braun came to my laboratory and proposed that we work on the possibility to have the Redstone missile used [to launch] a small satellite." No official authority for the program existed. The engineer oversaw a small design group in his laboratory, one that was not very visible. "That's probably why he came to us, because you could camouflage it much better."

The laboratory workers specialized in guidance and control—the problem of steering rockets on the proper trajectory. The workers came to the point in the rocket development where they could not proceed without testing their ideas. "Von Braun had a splendid idea. . . . We had to test the nose cone of the Jupiter." Although von Braun had no authority to put a satellite into space, he did have authorization to conduct missile nose cone reentry tests using Jupiter C, a Redstone rocket with two upper stages added on. Von Braun arranged a test in which the first two stages of the Jupiter would propel the nose cone to a high altitude; the third rocket stage would then turn downward and push the nose cone into the atmosphere at a high reentry velocity. Except for the last downward turn, the procedure was the same as the one needed to propel a satellite almost into space.

After one or two attempts, they had a perfect test. "I was out on an observation station with Wernher von Braun and he got the Doppler information." The Doppler test allowed the team to calculate the speed that the rocket had attained. It revealed that the rocket had reached sufficient velocity to put the nose cone into orbit, if only they had changed the trajectory and added a small fourth stage. The nose cone test had been a complete success, but von Braun did not seem pleased. "You should have put the [fourth] rocket on," von Braun said. "You would be in orbit."

"It was a nice experience for me," said the member of the German rocket team. Von Braun never lost sight of his ultimate goal. "He was always aiming at the next step."[90]

"We were really looking for what we were going to do that would put our names in the history books," said one scientist who left an industrial research laboratory to come to NASA. "Here was an opportunity to really make history and shape space science."[91] As a NASA executive who left a military career to join the civilian space program observed, "how many people in the world could lead a program to build a complex to launch men to go to the Moon?" Not many people have a desire to work for less pay than they could get elsewhere, he observed. It is hard to get exceptional people to work for the government "unless there is the compensation that you are doing something that can't be done anywhere else. . . . You've got to be doing something that is valuable. You've got to be doing something that is unique."[92]

It is hard to measure the objective reality behind this NASA belief. NASA employees, particularly those who served during the first decade, insist that the people who came to the agency were extraordinary. They point to cases of former NASA employees who went on to successful careers in industry or with other government agencies. Since the government does not collect statistics on the comparative intelligence of federal employees, the belief is hard to confirm.[93] The objective reality, moreover, was probably less important than the ways in which the belief affected behavior. Belief shaped practices in a number of important ways.

First of all, the belief supported a norm of hard work. "We just worked sixty, seventy hours a week for most of those years," said one of the founders of the space agency. Because NASA officials expected each other to be exceptional, they expected each other to work exceptional hours. In the beginning, the team of officials with which this person worked commuted to Washington. "We would go down Sunday and come back late Friday night or early Saturday morning. We would start at seven in the morning and break for dinner at night and work some more after dinner and then go back to [the] hotel and have a little bourbon and talk the day's problems over and finally quit around eleven o'clock. We really worked." It never let up, he said, during the ten years that he worked in Washington. "I probably worked between sixty and seventy hours a week, sometimes more. We all did it. That was part of the culture." This official recalled an organizational meeting his team had with Keith Glennan, NASA's first administrator. Glennan stood up and said two things. "He said, you have got to go at this space program as if you were fighting a war. And he said, this is no place for tired men."[94]

The norm of hard work was supported by the fact that NASA during its formative years developed a very youthful culture. "Do you know what the [average] age of my organization was in 1969 when we flew to the Moon?" asked one the directors of the flight operations division at the Johnson Space Center. "Twenty-six."[95]

"The space program is a young man's game," said another NASA executive. "It's hard work," he added. "You need to be at your peak."

He illustrated the point by telling a story about the person recruited to fill his job when he left. The executive chaired the search committee. At one point in the interview process, he said to the person who would be his successor: "This is a rough job. Do you think you've got the stamina to handle it?"

A year later he was talking to his successor and he said, "You know, you made me furious when you asked me that question." He said, "I really couldn't believe that you were asking me that question. I just felt you were stupid. Obviously, you didn't know that I was a mountain climber." The successor, in fact, climbed in the Himalayas. "You didn't know what kind of physical stamina I had." Then he said, "Here I am, only a year into the job, and I am already dragging ass. It's a damned hard job."[96]

The rapid expansion of the space program during the formative years, combined with the relative ease with which people could move in and out of NASA, helped to create an institution in which a disproportionate number of the employees were young. This is not untypical for an organization engaged in a crash program that must suddenly recruit large numbers of employees to carry out challenging tasks. By their nature, such organizations tend to attract large numbers of young people, or, if older, people who have not been with the agency for long periods of time. The youth culture of NASA's first decade was indeed a fact. The average age of scientists and engineers working for NASA during the mid-1960s was thirty-eight, the lowest average during its first three decades.[97]

The youth culture contributed to the flexibility of the agency and the can-do attitude that characterized the early space program. "The young guy," said one NASA leader who had been with the NACA since 1945, "doesn't know that the road is paved with all kinds of trouble, and politics, and money, and failures, and all that sort of thing. He can run down the damn road like crazy." Older people contribute experience, this official said, and help exuberant young people find their way around the pitfalls and shoals. "New ideas, new capabilities, new ways of doing things all come from the young. They don't come from old folks like me. What you get out of people like me is experience." At the time that Americans landed on the Moon, this senior official, one of NASA's top executives, was forty-five years of age.[98]

Because NASA employees believed they were special, they would not tolerate someone in a key post who was not performing well. "We lost lots of project managers," said one NASA official. "I don't want to put their names down because I remember one of them that came and sat on my side porch on a Sunday afternoon and cried. . . . We had studied every damn thing you can think of to figure out how to do this and he failed. He failed. There were several that failed at Houston. Subsystems people, major systems, and so forth. We had lots of them."[99]

The expectation of exceptional performance allowed agency officials to demand what became a central element of the NASA culture. "Painstaking attention to detail," wrote one leading NASA official, was the "one overriding consideration that stands out above all the others" in making possible the expeditions to the Moon.[100] Whenever NASA had a failure, the official added, "the reason was always the same: we had failed to be inquisitive." They had failed to follow the tendency of the engineer to become preoccupied with minutiae. "Remember *Apollo 13?*" the official asked. "The spaceship's oxygen tank exploded when we were halfway to the moon, and we almost lost three men. Why? Because years before, we had overlooked the obvious." In their obsession to eliminate fire hazards inside the spacecraft, they overlooked one important detail. "Nobody asked about the fire hazard inside a tank of liquid oxygen." (Oxygen tanks in the service module contained internal heaters, small fans, and controlling switches. A short in the electric wires inside the tank ignited Teflon insulation and caused the tank to explode as the spacecraft approached the Moon.) "That's exactly where a fire started."[101]

From a distance, space expeditions and rocket launches can seem like terribly exciting affairs. In fact, the job of preparing them is monstrously tedious. Preflight checkouts and paper trails become routine. When tedium makes for complacency, the groundwork is laid for disaster. The emphasis upon attention to detail was one of the principal means by which NASA officials could overcome that danger.

Lack of attention to detail could ruin even simple projects. "When I was in the Office of Aeronautics and Space Technology," said one NASA engineer, "we had a project . . . to simulate airplane landings and tires and brakes." The project made use of a landing loads track at the Langley Research Center. This was a long track along which a heavy sled was propelled, with the experiment attached to the sled. "It accelerated down this track to about 200 knots and then you do your experiment." Large hydraulic tubes stood at the end of the track. Through the physics of compression, they slowed the sled to a stop. "We had a nice checklist in the building." The person in charge of the track would run down the checklist before starting the test. "Pretty soon he got to know it really well, so he stopped using the checklist. He just did it automatically.

Workers inside the shed had left one of the hydraulic valves open. "So the guy came in and he went through the drill, didn't check anything," and he closed the open valve instead of opening it. Down the track came the sled, with no hydraulic pressure to stop it. "It went off the end of the track and through the building at the far end." It was a perfect example, said the engineer, "of what happens when things become so routine that you no longer need to check."[102]

Stories such as these gave credence to NASA's obsession with detail: the simplest things, once overlooked, created problems. One NASA official from the Apollo era remembered how he had to scrub a launch, "before God and the world, you know," because of a simple mistake. Workers had put a barrier in a cryogenic fuel line in order to run a test. (Cryogenic fuel is very cold fuel, such as liquid hydrogen at minus 423 degrees Fahrenheit.) "We didn't take [the barrier] out. So we went to flow cryogenics, and I couldn't get the cryo on the bird." He had to go out and tell the press that they could not launch because of a simple mistake.

NASA officials developed a number of techniques to keep track of simple details like these. "We learned," explained this NASA launch official, that "any time you break a system, any time you put something in as a replacement . . . you had better put on an identifier." For rockets being prepared for launch, they used streamers with red tags on them and a piece of paper that identified the part in question. "For example," he said, "like your pyrotechnics." NASA used explosive devices throughout the rocket and its spacecraft, such as explosive charges that separated the rocket stages during flight. "You can't put your pyrotechnics in until the last week before your flight, so you have to have something else screwed in there." Red streamers would be attached to the dummy part. Streamers would appear for something as simple as a missing bolt. "Sometimes when you go to lift these, the guys say, all I'm seeing is red tags. Christmas packages."[103] There would be red streamers everywhere.

NASA officials developed very detailed paper trails to keep track of program details. "You have got to track the whole process all the way through," said one headquarters official, "more than any other program that I have ever been associated with." The tracking began at the factory and ran all the way through the test and verification process to the launch. "Let me give you an example. . . . We had a big problem with the strength of aluminum in the very early days of the shuttle program." They discovered a piece of aluminum that had been improperly heat treated. The paper trail allowed them to trace the aluminum back to the manufacturer and then out again to where it was used. "Because of our records we knew all of the lots of the aluminum that we were using in the shuttle and took appropriate action."[104]

Attention to detail was reinforced through a variety of review meetings. "I used to have a technique called management by embarrassment," an Apollo launch official said. "We'd all sit together, like every second Tuesday; had everybody around the table, all my leaders, company and government." Each person would stand up and announce whether or not their work was on schedule and what problems they might be having. "No one wanted to be the laggard."

"One day I was at the pad," this official recalled, "and I found a red tag somewhere on the launcher." The wind had blown it off the rocket. "At the next meeting of the group, I had the tag there, and I said to all the contractors, I said, I've got a tag here with a number on it. It belongs to one of you. I want to see how good your system is. I want you to tell me who's missing this tag." He could tell whose tag it was, because of the number on the tag. "But the fact is now, did the guy know he's missing it? That was the check. Well, at the next meeting, all of them swore it was not theirs. And I called the leader of the team front and center and presented it to him, and it was his." Small details like that drove teams of people to seek perfection. "There's the embarrassment of coming up to receive, you may say, a medal that you don't want to receive."

NASA officials from the Apollo era insisted that the space program worked because exceptional people of special ability made it work: individual people mastered the details, then worked together in harmony. "That was not just in Florida," the launch official said. "That team goes back to the mechanic, whether you are at Bethpage working on the LEM, whether you are in Downey working on the command module, whether you are in Michoud working on the first stage. . . . It's something so unique. It's so—so beautiful. In many ways, it's like a Toscanini march. Everybody's got their job; everybody's there. But the old masters say, okay, here we go, and then they play. Without those guys having worked together, it's just a discordant bunch of noise."[105]

Exceptional people were able to overcome obstacles thrown up by the bureaucracy. "We made it work," said one of the engineers who led the Apollo expeditions. "You go back and look at the management of Apollo," he argued. "It worked because people like Gilruth, and Low, and Kraft, and Phillips, and others recognized the damn limitations of the management scheme, not because of it."[106] People made it work.

3 ▲ Breaking Barriers

The plain fact is that we are in a business that is
both hazardous and highly experimental.—James E. Webb, 1967

At the time that Congress established the National Aeronautics and Space Administration in the summer of 1958, the United States had put only four satellites into orbit around the earth. No humans had flown in space. No robotic probes had taken close-up pictures of the Moon or landed on the planets. NASA received a mandate to establish American preeminence in space.[1] Congress happily provided funds sufficient to carry out the mandate. The public viewed the space venture as a supreme test of the nation's technology and organizational skill, an appropriate response to an intense cold war challenge from the Soviet Union, whose space spectaculars had shaken confidence in American institutions. People expected NASA to break barriers, much as aeronautical experts had done eleven years earlier when they pushed airplanes through the sound barrier.

NASA engaged in a crash program, undertaking exploration adventures of the first magnitude. NASA executives enlisted in this effort a generation of engineers and scientists whose lives had been shaped by the experience of the Great Depression and the Second World War. Both the nature of the task that NASA faced and the generation of people gathered to carry it out, factors sharply fixed in time, shaped NASA's original culture.

Risk and Failure

NASA employees understood that they could not explore space without taking risks. They sought to minimize risks, but they could not eliminate them. To totally eliminate risk they had to cease exploring, an unacceptable option given their mandate. With risk came failure. It was rare for NASA to conduct a space flight mission on which something did not go wrong. Some level of failure was normal given the missions that NASA was expected to perform, especially during the early stages of a program. As part of their culture, NASA employees came to believe that risk and failure were normal. Ninety-seven percent of NASA professional employees surveyed in 1988 agreed that "risk and failure are a normal part of the business of developing new technologies" (see the appendix: attitudes toward risk and failure).

"You didn't learn except by failure," said one of the leaders of the Apollo program. "Now, you didn't set out to kill people and you didn't ever fly a machine in a flight regime where you didn't have a reasonably good understanding of what the flight characteristics were going to be or the environment that you were going to fly in. But—and this is a very callous statement—we of the flight test business were acquainted with death." NASA officials chose astronauts from the corps of aircraft test pilots for a deliberate reason, "and that deliberateness was that these men were used to putting their lives on the line every time they flew."[2]

Part of the legend of the test flight business during the 1950s was the belief that one out of four test pilots pushing the barriers of supersonic flight died at their work.[3] Between 1948 and 1967, three NASA and NACA test pilots died at Edwards Air Force Base in test aircraft.[4] More still died in the military flight test programs.[5] "That was a characteristic we had to have," said the Apollo program leader. "Anybody that gets on the end of a flaming rocket and doesn't recognize the risks and dangers associated with it, does not understand the problem. We were well aware of the risks we were taking. On the other hand—and I emphasize this very, very carefully—we would never fly a manned vehicle if we knew something was wrong with it until we fixed it. That isn't to say that there weren't some unknowns. That isn't to say that we didn't recognize the risks involved in the operation every damn time we went to the pad." He concluded with a frequently made observation: "Recognition of risk is what made us as good as we were."[6]

"You knew how risky it was," said another Apollo expedition leader. "We certainly knew the risk, my goodness, yes. We were all airplane [test] people, we knew how many times they crashed." They did what they could to reduce risks, but they were "not going to put any blindfolds over anybody" and deny that the risk was not there. "If there is a risk you are going to say, this is a big risk and we've got to do whatever we can." They would not take the risk if they knew that they could not get through it. "But if it was, like, one in one hundred, you would do it, you would take it. . . . There were so many ways it could happen. It would be very bad to have a man in orbit and not have the retrofire work and you would just leave him up there and he would die in orbit and say goodbye to his friends and his people and tear everybody's heart out. Now, that would have been a bad one."[7]

Risk came in many forms. NASA pushed the Mercury astronauts into orbit on a modified Atlas intercontinental ballistic missile. The astronaut lay in a capsule that NASA officials substituted for the missile nose cone. "The Atlas," one NASA official observed, "had been designed as a ballistic missile. It was never assumed that it would be 100 percent successful."[8] The Air Force, for whom the missile was originally designed, could accept a less than perfect reliability rate because it was less expensive to buy a few extra missiles to make up for the ones that might fail than to stretch the reliability. In the first trial of the Atlas-Mercury system, the Atlas rocket tore apart one minute after lift-off. Fortunately, it was an unmanned test. The cause was never identified.[9]

"We knew that we were going to have a lot of failures," said the NASA official. "I used to refer to our attempts to launch as random successes. That was okay at the time, because we were forging such brand new ground. . . . People expected people to make mistakes and we didn't have enough knowledge about what we were doing to avoid taking risks."[10] A top manned space flight official remembered "one protagonist claiming that we were going to kill the astronauts in our Gemini flights by allowing them to stay up there more than a day—and going to Congress with a whole set of letters saying that."[11] NASA made it through ten Gemini flights, but not without some heart-stopping moments.

The second test flight of the Saturn V rocket—the launch vehicle designed to propel American astronauts toward the Moon—failed to work as planned. It was a test flight with no astronauts on board. The whole vehicle, including the first-stage F-1 engines, vibrated during flight. When the vibration frequencies began to harmonize, the rocket started to oscillate like a giant pogo stick. "We had pogo," one top official recalled, "and had one or more J-2 engines fail on the second stage." A fuel line leading to one of the J-2 engines ruptured, spraying the upper section of the engine with liquid hydrogen before the engine shut down. Investigators located the problem and devised solutions. The very next Saturn V rocket to fly launched a three-person crew on a flight around the Moon.[12] Had the rocket failed, NASA would have been subject to intense criticism. "But, in a carefully assessed risk sense," said the official, "it was a prudent move." From a purely technical perspective, there was no reason to wait. "It was done with great precision," he added.[13]

Unavoidable risks accompanied not only the flights with humans on board, but unmanned programs as well. The technological challenge of operating automated spacecraft thousands of miles from the earth, when joined with the ever-present possibility of human error, created frequent opportunities for full-scale failure. In the early 1960s, NASA developed the Ranger program, an automated effort to provide close-up pictures of the Moon. The rocket engine for *Ranger 1*, launched in August 1961, failed to restart in space. The attitude control system on the launch vehicle for *Ranger 2* put the second probe into the wrong orbit. The next four Rangers failed as well.[14] Congress launched an investigation, changes were made, and the program continued.[15] "Failure was more acceptable then," said one of NASA's science directors. "It was really part of the business."[16] Not until the summer of 1964, with the launch of *Ranger 7*, did NASA carry out a successful Ranger mission.

The Surveyor mission was one of the riskiest that NASA undertook. Surveyor was an automated probe designed to precede the astronauts to the surface of the Moon. "The Surveyor mission," NASA officials wrote, had "very little margin for error." The radar altimeter Doppler velocity sensor, the on-board device that controlled the descent to the Moon, was a unique piece of machinery. "Nothing of this type had ever been developed before," yet it had to work perfectly for the probe to land. NASA officials told their congressional authorization committee, "Unless virtually all equipment functions within specifications, the flight will fail."[17] On June 2, 1966, the first Surveyor landed on the lunar surface. Six more followed; two of them failed during their descent and crashed on the Moon.

Shortly after the founding of the agency, NASA officials developed the concept of a general purpose scientific satellite to be flown with a variety of instruments on board. Solid-state transistors were a relatively new technology at the time. One of the science directors recalled how some of the contractor bids proposed that the satellite use vacuum tubes, an established technology. TRW, which became the prime contractor, proposed the use of transistors. "I remember the head of reliability at the time back then saying, you know, we can't take that chance." Transistors were too risky. "Well, I said, you know we are supposed to be in the research and development business. We are supposed to be on the cutting edge." The agency used transistors. The first Orbiting Geophysical Observatory, launched in 1964, failed in orbit when its boom deployed incorrectly. So did the second, launched in 1965, due to a launch vehicle guidance problem. "It was inherent in the job to take risks," the science director observed.[18]

NASA officials embraced many norms and practices in their effort to reduce failures. Probably the most important was their attitude toward failure. As contradictory as this may seem, the anticipation of failure led to its avoidance. "You have to seek out trouble," said one top headquarters official from the Apollo era. "You had to make trouble your best friend." Initially, he said, even though a project seemed to go well, experienced space flight personnel knew that trouble lay ahead. "After you have been in the game awhile you begin to wait for the trouble to arise. You know that you are going to have it and you want to get into it as early as you can. Until you have had [trouble], gotten into it, and solved it . . . you are not really sure that your project will even work. . . . There used to be, starting from the top, from the Administrator down, very much of a concern to look for trouble, to try to anticipate what might go wrong in a project. You were expected to say 'what if'—what if this went wrong, what if that went wrong."[19] NASA executives did not countenance the bureaucratic tendency to avoid failure by denying that it could happen. Only by accepting the reality of failure could the organization deal effectively with it.

The normalization of risk, the acceptance of failure, and the anticipation of trouble led to an atmosphere in which these things could be discussed openly. NASA's ability to handle risk required open discussions in which mid-level managers and engineers felt unrestrained in voicing warnings and dissent.[20] "There was no tolerance for the 'yes' man," one top NASA official observed.[21] Managers relied on the opposition speaking out so that people could weigh all options. "If you thought something was a problem or if you had an idea on how to change something, you could bring it up and it would get talked about and solved," another NASA executive added.[22] Open communication was very much a part of the early NASA culture.

"There was a great deal of democracy in the management," said one of the agency's top spacecraft engineers. "Everybody . . . was free to state their feelings. No one was treated any different if he objected to what management would think than if he praised what management would think. Management didn't look for praise. They looked for anybody with good advice." This attitude had its roots in the technical cultures of NASA's predecessor organizations. "Management in the

NACA," the engineer continued, "depended on the staff, the working level staff that was doing the research, the guy in the trenches to come up with ideas." They shied away from "the heavy-handed approach of saying, okay, I've decided what we're going to do and you do that and you do that." Democratic management and a spirit of open communication, however, did not mean that everyone got along with everyone else. People fought hard over their ideas. "There were a lot of very strong personalities there," the engineer observed. It required a strong leader to make sure that everyone had a chance to speak and to make people listen to the opposition.[23]

As members of an engineering organization, it seemed natural that NASA officials would design machinery that improved the capacity for open communications. One of the top officials at the Kennedy Space Center described how they engineered communication loops. "I think we had 250 channels [on which] people could talk together in Complex 39," the entire facility for launching astronauts on their journey to the Moon. The Boeing Company, which oversaw the ground support equipment for the Saturn V, "you might give them twelve channels. They in turn would break it out. Mechanical would be one, electrical two, this three, instrumentation four." Ten people might work on one channel. They could talk individually, or everyone could talk. "In turn, my government guy working with Boeing, he could get any one of the ten. I could get them all, and you worked a progression." NASA launch officials would check with the various people up and down the communication line. "You could tune into North American 2, and you'd be listening to the guys working the engine." (North American Aviation assembled the Saturn V second stage.) "If there was a problem there, you could hear how they were handling the problem." When the time came for the launch, they put all of the key people on one loop, and each person checked in. "You got instantaneous communication up and down. That discipline, just on that loop and how they handled themselves, tells you a lot. [It was] probably one of the biggest loops ever put together. The aircraft carriers have things like that, but I don't know how many channels they have. It's instantaneous communication, instantaneous transmission of knowledge."[24]

NASA officials developed management review procedures to enhance communication. One NASA executive recalled how they reviewed their space science projects. Once each month officials would critique their projects, starting with the contractors and working up. "You classified a project green or yellow or red. If it was green, it was in good shape. If it was colored yellow, you knew you had better watch it closely, trouble was coming. If it was colored red, you knew you were in real trouble and you had better be working on a solution." The review eventually wound its way to the office of the NASA Administrator. "The problems surfaced; they rapidly moved up to the level that they needed to be at in order to be resolved, decisions were made to take the action necessary to resolve the problem, and that went back down." The executive characterized the process as a "taut" communications network.

"A classic example of a management communications system that was not taut was the case of the Viking biology package."[25] Biology packages were an important element on the two Viking probes that landed on Mars in 1976. An automated arm on the Viking landers scooped up a sample of the Martian soil and placed it in an automated biology laboratory on top of the spacecraft that tested the soil for signs of life. The TRW Corporation received the contract to build the biology package.[26] "In the early days of that project, we had what we called a mail drop approach," said the same executive. The project manager at TRW would send a letter to the Martin Marietta Company, the prime contractor for the lander, informing Martin Marietta of a problem. The prime contractor would in turn send a letter to the project director at the Langley Research Center. "A letter would go back and down and ultimately something might happen. But by then usually another problem had erupted and we had lost another month or two out of the schedule."

"Well, we discovered that we really were in deep yogurt on Viking in late '74 and '75, and the way we then solved that problem was that once a month I would go out to Los Angeles to TRW. The director of Langley would go, the project manager would go, and the CEO of Martin Marietta would attend." All of the responsible officials would gather in one room. "The project manager and all his system managers would report on the status of the Viking biology package, including the scientific experiments. At the beginning, we had like ten insoluble problems. By the second meeting we had twenty, and ultimately by about the fourth meeting we had fifty problems. Along about the fifth or sixth meeting, the number of insoluble problems began to drop. Ultimately we solved all the problems." The biology package flew to Mars on the spacecraft, searched for life, and found none.

"The reason that it worked and that we got it ready on schedule was because we had everybody in that room that we needed to make a decision." The scientists, the project team, the contractors, the people who could commit funds were all there. "It got to the point where we could identify a problem in the morning and by the close of business we could solve it, get the money allocated, get the decisions made and get things working. Now that was what I would call a taut management system."[27]

NASA officials set up committees that pried into the work of project managers, an administrative equivalent of the checks and balances that regulated the government at large. "I called it a murder committee," said the same executive. "There were about ten or twelve people on it from the other centers who had nothing to do with Viking. If it dropped into the drink or blew up, it was no skin off their nose." NASA executives brought in the committee members and told them to sit in on the management reviews and ask all the questions they wanted to. "Tell us if there are things that you see that worry you," they were told. "We want you to first voice your concerns to the project manager. If you have concerns about the project manager, voice them to us, and if you don't get the satisfaction that you want, then voice your concerns to the Administrator of NASA."

The members of the committee were willing to participate because they knew their recommendations would be heard. "The project manager was comfortable with them," the executive proffered, "because they were there first of all to provide advice and help to him. They were not [individuals] who were looking over his shoulder, so he had to cover up what he was doing."[28] NASA officials used the same system of oversight committees to check the work of people running the Apollo program.[29]

NASA officials developed machinery that reduced the risk of catastrophic failure. They placed a rocket on top of the Mercury space capsule, with sensors that could gauge the performance of the Atlas launch vehicle. If anything went wrong, the escape rocket was designed to jerk the capsule and its astronaut off the top of the Atlas and away from the launch pad. NASA used the same system on its Gemini and Apollo flights. Planetary probes, from Viking to Voyager, went out in pairs. "If one of them crapped out," said NASA's science director, "you wanted another shot at it."[30]

Flight directors at Mission Control developed rules to protect the lives of the astronauts. The rules required flight controllers to balance one risk against another.[31] Two hours into the flight of *Gemini V*, a planned eight-day voyage, the fuel cell that produced electricity for the spacecraft started to fail. The crew prepared for reentry. The dangers of leaving the crew in orbit to perform their experiments, however, were less serious than the dangers of subjecting a new crew to another launch. Flight directors decided to continue the mission, accepting the lesser risks posed by the balky fuel cell.[32]

NASA officials sensed that people to whom they were ultimately accountable understood the inevitability of failure. "We didn't have enough knowledge about what we were doing to avoid taking risks," said one top NASA executive. "Mistakes were a much more accepted kind of thing than they are today." In their first full year of operation (1959), NASA officials conducted fourteen test flights and satellite launches. Six of them failed, most from rocket malfunctions.[33] "The outside world was quite willing to accept the fact that we had failure rates which we would not tolerate today."[34]

On January 27, 1967, technicians at Launch Complex 34 preparing for Apollo-Saturn mission 204 powered up an Apollo spacecraft for a full-scale ground test. The actual mission, scheduled for later that year, would give the Apollo spacecraft its first orbital workout with astronauts on board. The spacecraft with three astronauts on board sat atop a Saturn IB rocket. An army of technicians and NASA controllers prepared to simulate the launch of the Saturn IB. Earth-based atmosphere was pumped out of the spacecraft; flight atmosphere of 100 percent oxygen was pumped in. A fire started somewhere in the capsule. It spread quickly in the oxygen-rich atmosphere, killing the three astronauts before technicians could break in.

"That put an indelible mark in our brains," said one of the NASA officials responsible for the test. The deaths of Virgil Grissom, Roger Chaffee, and Edward White marked a turning point in NASA's perception of the tolerance for risks and errors. "I never thought the reaction would be quite so negative and severe," a launch official noted.[35] Investigations and recriminations followed. Congress increased its oversight of the civilian space program. NASA replaced the Apollo spacecraft program manager and tightened up its program management style.

The NASA official sent to Houston to take charge of the Apollo program reported that his predecessor "had followed the policy in Houston of obtaining the very best men they could for the senior positions and had, as a part of the process of obtaining them, given assurances that they would have almost complete freedom in carrying out their responsibilities." This approach was inadequate, wrote the NASA Administrator, who ordered that a tighter system of supervision be established.[36]

"The fire did give us a baptism," the launch official observed. "We knew from then on there would be no forgiveness." NASA employees continued to believe in the inevitability of risk and failure. That remained part of the NASA culture. Through the aftermath of the Apollo fire, however, they began to understand that neither the public nor their elected representatives would tolerate catastrophic failures that could be traced to human error, as most can. Another catastrophic failure, with the loss of astronauts' lives, would have devastated the Apollo flight program. "There would be no forgiveness in this country or in the press or understanding of what we were attempting to do," the launch official said.[37]

"There is a list of all the times that we went to the Moon," said one of the NASA engineers who organized the expeditions. "Finally I said, look, this is the last time. We are not going to go again, because you are just asking for trouble." The last Apollo expedition left the surface of the Moon on December 14, 1972. "It's a very, very risky thing to do."[38]

Even as the reality of limited outside tolerance remained, NASA officials told stories reaffirming their belief that employees should be allowed to fail. One story in particular took on mythical qualities, the historical basis for the tale becoming blurred while the moral lesson remained clear. "This was on the Redstone missile," said one of the German engineers recalling the tale. The Redstone was developed by the German rocket team as a battlefield missile; it was later used to launch astronauts Alan Shepard and Virgil Grissom on two suborbital flights. The incident took place during one of the many flight tests of the Redstone rocket. On this particular test, the rocket failed. "It failed in about the twenty-third second." At the time of the failure, there was an unexpected shift in the roll command guiding the movement of the rocket.[39] "Now during the acceptance test, before launch, everything was in perfect condition," said the German rocketeer. "And I said to myself, how could this have happened?" Try as it could, the rocket team could not find the cause of the failure.

The German engineer continued. "I went to a responsible man in my laboratory and said . . . could it have been that this is a potentiometer which has a very close wiring so that the center pad shifted due to vibrations?" Misalignment of the potentiometer, part of the mechanism that controlled the flight of the rocket, would result in a feedback error that could account for the errant turn. "And the fellow said, 'you are right . . . I wanted just to see you this morning.'" After the final checkout of the instruments, he had tightened the screws on the center pad, inadvertently changing the setting on the potentiometer. "'That's the way it must have happened,'" the man admitted. "I said, let's go right away to von Braun and report it."[40]

One of von Braun's assistants continued the story. "We had always maintained that if somebody feels something is wrong, to come up and say something. And not just hide it, but rather come out into the open with it so that we could get in and correct it, which I believe is a very, very good thing to do." In a culture that accepts risk and failure, people are encouraged to bring forward problems and errors. Such an attitude, *and rewarded* the assistant said, "was the only way to really come up with outstanding work."

"Von Braun had two possibilities. Either give him a bottle of champagne or fire him." As a result of this man's error, a rocket had been destroyed. "Two possibilities. Either fire the guy, or shake his hand and say, 'It's very fine that you admitted you made that mistake, and now let's correct it.' . . . So von Braun said, 'Okay, this is good that you said this rather than hiding it, and taking some chances.'" Von Braun praised the man and brought him a bottle of champagne that night.[41]

The story spread throughout the agency. The engineer who received the champagne later maintained that someone else misaligned the instrument, and von Braun gave the engineer the champagne only because he solved the problem.[42] By then, however, the point of the story had been registered, and people continued to tell the tale. "If you made a mistake, you made it only one time," added another of von Braun's team. "The same mistake, you are allowed one time."[43]

Frontier Mentality

Engineers and scientists dominate NASA. They outnumber all other occupational groups. Nearly all of the people who accede to the top management positions are engineers or scientists. Within NASA, engineers like to engineer. They are most happy when they are designing things, building models, testing them to see how they work, perfecting designs, then moving on to new challenges. NASA scientists like to do science. They are most content when they are preparing experiments and making new discoveries. Only about 14 percent of NASA's engineers and scientists leave their work and go into full-time management.[44]

The desire of NASA engineers and scientists to move on to new challenges and make new discoveries neatly matched the nature of space exploration during NASA's formative years. In the human space flight program, NASA officials graduated through a series of distinct challenges. Once they completed the Mercury program, they moved on to the Gemini program. Once they mastered the Gemini flight program, they moved on to Apollo. Each flight series presented new challenges that, when completed, led to higher plateaus. This was also true for rocket, satellite, and space probe development. After three Mariner spacecraft flew by the planet Mars, *Mariner 9* went into a Martian orbit. After *Mariner 9* mapped the whole planet, NASA officials sent *Viking 1* and *Viking 2* to the surface of Mars to look for life. It was remembered as a golden era by those who worked for NASA.

Conversely, NASA engineers and scientists tend to be less content when they are stuck on the routine, especially when that routine requires them to push paper or supervise contractors. The professional work force in NASA tends to view time spent on management as time taken away from basic engineering and science, an observation confirmed by the 1988 culture survey (see appendix: attitudes toward management and change). Although many NASA engineers pursue management responsibilities—management jobs remain the primary route up the career and status ladder in this as in other government agencies—a significant number of NASA professionals believe that engineers do not make good managers. (As the culture survey results reveal, they believe that scientists do even worse.) Nor are these professionals content running established programs. Once a program becomes routine, they would just as soon move on to something else. By a margin of three to one, they disagree with the statement that "NASA should concentrate on implementing programs already approved rather than pushing for new programs." The frontier mentality is very much a part of NASA's technical culture.

"We were a typical, pretty well-organized research and development organization," observed one of the leaders from the 1960s. "We were not an operational agency, and we never pretended we were." Operational agencies concentrate on mastering routine. NASA officials concentrated on new frontiers. When a NASA program moved out of the research and development phase and became operational, the dominant philosophy required that it be spun off to another agency. Determining exactly where the dividing line between R&D and operational status lay could be a difficult task.

"The best example is the weather satellites," the official observed. "We had a yelling and screaming session several days over that." James Webb, NASA's second administrator, did not want NASA to get into the business of operating meteorological satellites. That responsibility rested with the Weather Bureau in the Department of Commerce. NASA, however, was responsible for developing the technology for each new generation of weather satellites and launching the satellites once they were built. NASA engineers wanted to keep improving the satellite design. Webb insisted that the program move to the Weather Bureau. "Webb says, we are done on this. Yes, it may need some little bit more research and development. We can do that for the Department of Commerce. We are going to transfer it over. And everybody said, 'No, we are not. This thing is not ready. It is still R&D.'" The official went to the project director and told him to let the program go. "I'm tired of this nonsense," the official said. "It is going to the Department of Commerce." The Weather Bureau took over the operation of the Tiros weather satellite program in 1965. "And it worked fine."[45]

"The cult and culture of the engineer," said one of NASA's leading space scientists, "is invent and build, make it work, watch it work and then go up and invent, build and watch something else work." Staying with the same design over and over again was not part of NASA's original culture. "Designing something and getting it to operate smoothly, nicely, so you can use it time and time again forever is not something that gives a great number of our engineers the jollies."[46]

During the formative years, officials fulfilled this part of the NASA culture by giving their people new things to do. "We were always doing something harder each time," said one of the leaders of the early space flight program. "The first time we put a man into orbit, the next time we put him into an orbit that lasted a lot longer, then we made a rendezvous one thing with another, and so on and so forth. Our program was an unfolding one. It was unfolding all the time into more difficult things, ending up with flying men to the moon, which God knows was a tremendous thing to do."[47] New challenges kept people alert and prevented complacency.

In the earliest days of the space program, the desire to pursue new frontiers provided justification for an important aspect of NASA's culture. Not only did NASA officials ascend a series of technological and scientific plateaus in their approved programs, they also became advocates for space endeavors not yet approved. The original 1958 legislation setting up the civilian space agency gave NASA a vague mandate to establish "long-range studies" on the opportunities for space exploration and preserve the role of the United States "as a leader" in aeronautical and space technology.[48] The attitudes of NASA engineers and scientists amplified this mandate. "Engineers have a special responsibility to be bold and imaginative," NACA Director Hugh Dryden said of his agency's employees in a speech setting out a long-range vision for the nation's space program eight months before NASA was officially formed. "In my opinion, the goal of the program should be the development of manned satellites and the travel of man to the moon and nearby planets."[49]

Dryden became NASA's Deputy Administrator once it was formed. He and other NASA officials pushed hard for approval of a long-range plan that would carry out their vision. A Research Steering Committee on Manned Space Flight, made up of representatives from throughout the organization and formed in the spring of 1959, set out a long-range plan that included a space station, a lunar landing, and the exploration of Mars and Venus. George Low, NASA's chief of manned space flight, "recommended that the committee adopt the lunar landing mission as its present long range objective . . . because this approach will be easier to sell."[50] The objectives reappeared in an unofficial long-range plan issued at the end of 1959 by NASA's Office of Program Planning and Evaluation.[51] As one analyst observed, "NASA planners chose a lunar landing objective fully two years before President Kennedy announced his choice of the lunar landing as a national goal."[52] From the beginning, NASA officials saw themselves as advocates for missions that went considerably beyond what the President and Congress had actually authorized.

Dryden, Low, and other NASA career executives realized that the pursuit of these long-range plans would create technological challenges that would not only be difficult to solve but impossible to foresee. "No person can foresee the aeronautical and space activities of the future," Dryden claimed, "any more than the Wrights could foresee the aeronautical activities of today in 1903."[53] Some of NASA's earliest visions of space technology seem quaint by modern standards. In the 1960s, for example, NASA officials competed with a number of public and private organizations to develop the technology necessary to create a global network of communication satellites. NASA planners envisioned a fleet of ten to twenty plastic balloons, one hundred feet in diameter and coated with aluminum, to which radio signals would be directed and bounced back to distant points on earth.[54] NASA actually launched two Echo balloon satellites. Communication satellite technology progressed rapidly, leading to the fabrication of multiple-circuit communication satellites capable of simultaneously receiving and sending hundreds of signals. By 1969, with three such satellites parked in orbits over the Atlantic, Pacific, and Indian oceans, humans acquired their first global communications network.

NASA engineers and scientists spoke with excitement of the challenges of the early days. "When President Kennedy said we were going to go to the Moon by the end of this decade," recalled one of the engineers responsible for the first human flights into space, "most of us in the Space Task Group thought the guy was daft. I mean, we didn't think we could do it. We didn't refuse to accept the challenge, but God, we didn't know how to do [earth] orbit determination, much less project orbits to the Moon."

Simple problems, like moving fluids around in a spacecraft, had to be solved. "How do you get liquid out of a tank at zero-G?" the engineer asked. "Everybody said, 'Oh, what's so hard about that? You just pressurize it.' Pressurize it, my ass. The pressure exerts on everything, and the damn liquid is just floating around. It's liable to be in globs someplace in the tank, and you don't even know where it is in the tank. You don't even know how much you have got left in the tank." A technology for moving liquids out of tanks in microgravity was soon developed. "I mean, that's a fundamental now of zero-G. Everybody knows that you've got to put a bladder in there, or you've got to use capillary action to suck it out of there, or some other technique. But we didn't know that in 1958 and '59."

All sorts of new challenges confronted the first generation of NASA space flight engineers and scientists. "We didn't know what the condition of the lunar surface was. We didn't know what the radiation was going to be between here and the Moon. We had never heard of a fuel cell in 1960 and '61." The fuel cell, a complex technology that employs liquid hydrogen and oxygen to generate electricity, was developed to satisfy power requirements on long space flights. Development of fuel cell technology turned out to be a major technological challenge.[55]

New technologies, while compatible with the desire of NASA engineers to break new frontiers, also created new risks. "You pushed the frontier because that helped you get good people," said one of NASA's chief scientists. "If you are pushing the frontier, then people tend to be a little more on their toes." This was especially true when NASA rolled out a new or upgraded launch vehicle. "Everybody tends to be there and to worry about it and to make sure that it works."

He offered as an example improvements made to the Delta launch vehicle, the workhorse for NASA's satellite program. NASA officials used the Delta rocket (or Thor-Delta, as earlier versions were known) sixty-nine times between 1960 and 1969 to deliver payloads into space. "We started out with a certain capability in the early sixties, and by the early seventies we had improved its weight-carrying capability by about a factor of ten. One year we might improve the guidance, the next year we might substitute a new upper stage. There were always some changes to the system under way, so that the engineers were constantly challenged and had to keep on their toes." The descendant of Thor-Delta that could place 60 kilograms in orbit in 1960 could deliver 700 kilograms by 1974.

Reflecting on the risks involved, the scientist observed: "Sometimes, you got yourself out on a kind of a limb by doing that. I got myself out on quite a limb when we launched the Earth Resources Technology Satellite." NASA scheduled the first Earth Resources Technology Satellite, later renamed LANDSAT 1, for launch in 1972. The satellite was placed on top of the first Delta rocket to use nine solid rocket boosters strapped around the base of the first stage for extra thrust; the second stage had a newly uprated engine. "When we finally got ready to launch this multimillion-dollar satellite, I realized that it was going to be the first launch of what was almost a completely new rocket. The gods have to smile on you occasionally in these things."[56]

To reduce those risks, NASA professionals used proven technologies where possible. "We didn't try to invent new technologies for the sake of inventing new technologies. If we could see sufficient merit in the development of new technology we would certainly do it. We did it to reduce weight. We did it to reduce response time. We did it for a lot of different reasons, in some areas because technology didn't exist."[57]

NASA engineers and scientists acquired a considerable reputation for their spin-offs from space, earthbound applications of space technology.[58] Such spin-offs did occur, but much of the technology employed in NASA's early rockets and spacecraft was secondhand. Some people went so far as to suggest that the Apollo expeditions to the Moon required no major innovations or breakthroughs. The technologies used in the lunar expeditions, they said, had already been developed.[59] The truth appears to fall somewhere in between. When asked whether NASA in general relied upon existing technologies or pushed the state of the art, most persons interviewed for this study gave a uniform reply. NASA did both.

The German rocket team and other leaders at NASA's Huntsville center mixed old and new technologies. "They were the world's greatest incrementalists that I have ever seen in my life," said one headquarters official who watched the German team at work.[60] The first Saturn rockets employed tanks removed from the Redstone and Jupiter launch vehicles. The Jupiter was derived from the Redstone, which in turn was a second-generation German V-2. The engine technology on the Saturn first stage borrowed liberally from hardware built for the Jupiter, Thor, and Atlas missile programs.[61]

The J-2 engine for the Saturn V second and third stages, on the other hand, posed a much different challenge. The J-2 engine burned liquid hydrogen. Although ground-breaking work had been done on hydrogen-burning rockets like the Centaur, no one had ever built hydrogen-burning engines that could propel humans into space, nor ones as powerful as the J-2.[62] The J-2 engine also had to stop and restart in space. Without the engineering breakthroughs achieved on the J-2, it is doubtful that Americans could have reached the Moon.

Even a spacecraft as unique as the module that landed on the Moon mixed the old with the new. The on-board radar, employed for lunar landings and for rendezvous with the command module after liftoff from the Moon, required no technological breakthroughs. The ascent engine, whose infallible performance brought the astronauts up from the lunar surface, NASA professionals kept as simple as possible. It borrowed liberally from the Air Force Agena program, an upper-stage rocket powered by hypergolic propellants that self-ignited upon contact. The descent stage of the lunar module, on the other hand, posed a major technological challenge. Although the descent engine also used hypergolic propellants, it incorporated a complex throttling system that allowed the astronauts to control engine thrust during landing, and gimbals that allowed the engine to swing. Early versions had a discomforting tendency to chug.[63]

"If you want to make progress," said one of NASA's top engineers, "you've got to design things that have not been done before. And if they have never been done before, you can't guarantee they are going to work."[64] NASA's frontier culture encouraged agency employees to do things that had not been done before. It encouraged them to act as advocates within the government for the approval of new programs. Sometimes it meant taking risks and developing a technology in directions that had not been taken before. Sometimes it meant doing something new but doing it with an old technology. The latter strategy reduced risk and helped convince politicians of the feasibility of a new venture. The former strategy made new ventures possible. Altogether, the frontier mentality motivated NASA civil servants by appealing to their natural desire to move forward and improve.

The institutions that were combined to create the civilian space agency shaped NASA's early culture. So did the events of the early years. Institutions and events, however, were not the only forces at work. The people who enlisted in the American space program also forged the agency culture. Those people belonged to a generation that brought common values to their work, values formed in the families in which they grew up and in the experience of the Great Depression and the Second World War. These values defined the character of the first generation of space flight scientists and engineers. Members of the first generation were not readily disposed to compromise their behavior where such defining characteristics were involved.

Of all the means by which people in an organization establish a common culture, few are as powerful as the recruitment of employees who fit the organizational mold. People whose values coincide with those of the prevailing culture do not need breaking in or training in the cultural norms. In NASA's case, not only were the first-generation employees predisposed to behave in ways compatible with the agency's dominant cultural norms, their common values in many ways created those norms. Their life experiences had taught them the value of hard work and honesty and the importance of a good, technical education. The factors that attracted them to professions like engineering and science predisposed them to treat space flight problems as technical matters to be resolved through the application of strictly professional criteria. They were not prepared by upbringing and education to engage in political compromise where technology was involved. The space program also gave these individuals an opportunity to exercise the idealism to which they had been exposed in their formative years. The space race was "the good war," a challenge of public service undertaken to prove the superiority of the system from which they had emerged.

The first decade of space flight provided an ideal setting for the exercise of this orientation. The flight to the Moon and the other great exploration programs that NASA carried out during its formative years were as much a monument to the dominant middle-class values of truth and service as they were a technological achievement. Engineers were allowed to be engineers. They worked out technical solutions to the problems of space exploration, then dealt with whatever bureaucratic or political problems had been created by those solutions. During the decades that followed the landing on the Moon, this became harder to do. Political and bureaucratic considerations became more intrusive.

NASA officials continued to recruit people with the same values, but conditions surrounding their work changed. NASA professionals found themselves operating in a world where their basic outlook became more and more anachronistic.

Over half the members of the first generation of NASA employees were the sons and daughters of working-class families. As the 1988 culture survey reveals, over half had to work their way through college. Over 60 percent grew up in working-class neighborhoods, small towns, or in the country (see the appendix: origins). Only 2 percent were the sons and daughters of physicians, attorneys, or business executives. One of NASA's top scientists grew up on a small ranch in southeast Montana. "It was a marginal operation," he recalled. Thirty families had settled along Box Elder Creek on land that could possibly support three. "People tried to ranch with a thousand acres where they really needed ten thousand." The Depression and the drought arrived as he attended a local one-room school.

"My mother, being a school teacher, she read to me when I was young," he said, recalling the values his parents tried to instill in him when he was a child. "You didn't lie. You worked hard. You had chores to do." He tried to instill those values in his children, even though they grew up in Washington, D.C. "I always felt kind of bad that I brought

Table 1 Parent's Occupation, NASA Professional Employees

| | Year Joined NASA | |
| | 1951–1969 | 1970–1988 |
Occupation	(%)	(%)
Business executive, physician, attorney	2	5
Engineer, scientist, college professor, government executive, military officer	12	24
Salesman, accountant, bank officer, insurance officer, schoolteacher	17	21
Owner of a small business	16	9
Mechanic, electrician, plumber, carpenter, railroad engineer, machine operator, factory worker, postal worker, milkman, farmer, rancher, noncommissioned military officer, policeman, firefighter, nurse, secretary, bookkeeper, cashier, cook	54	42

Source: Author's 1988 NASA Culture Survey.
Note: The question asked respondents to check the job that most closely described their father's occupation while respondents were in high school. Respondents who were not raised by their father were asked to check the occupation of the head of the household in which they were raised while they were in high school. Data are statistically significant at the .01 level.

my kids into Washington, D.C., and raised them in an urban environment. I felt bad that they didn't have the kind of childhood that I had." Memories of his childhood were brought back to him one day at a briefing on the use of ATS-6, a communications satellite. NASA developed the Applications Technology Satellite (ATS) program in the early 1960s to test a variety of new technologies, such as collecting weather data and beaming radio and television signals down to the earth. ATS-6, launched in 1974, was designed to beam public health and education programs to small receivers in depressed areas in India and the United States.

During the briefing, the people making the presentation began to list the depressed areas to which the programs would be beamed. His childhood community in southeast Montana was on the list. "I was utterly horrified to discover that they considered this area in Montana and South Dakota where I was raised such a deprived area." The scientist did not believe that he had led a deprived childhood. "I felt that I had a much better opportunity than what my kids did here [in Washington, D.C.]."[65]

"Honesty was certainly high or at the top of the list," replied one of NASA's top Apollo program executives when asked about values. His father took a job as a lineman for an electric power company in order to support the family. "All of his adult life, he had a very strong desire to be a Baptist minister. At some point when I was still very young, he qualified for a license in the Baptist system. He had not been through college, and therefore . . . he didn't qualify in their criteria to be an ordained minister. But he was licensed."

"He spent most Sundays during my growing-up years as a voluntary minister to the country churches around Cheyenne. I got to know quite a bit of the ranch country out to the east and north of Cheyenne" (where he traveled with his father on Sundays during the 1920s and 1930s). "My father was a hard-working man. So was my mother for that matter," he added. "The economy was not all that good after the Big Crash. So I was well aware that my parents had to work hard to make ends meet with a large family. But that all worked out."[66]

"Values," explained one of NASA's top aeronautical engineers, "they were quite different from what are instilled in kids today." This engineer grew up in southwestern Virginia, where his father helped run a furniture manufacturing company. "I always recognized that I had to make a living and that I couldn't depend on anybody else to take care of me. That's not always true today."

"My father thought honesty was right above just about anything.

He could not stand any type of dishonesty." The engineer attended high school and the University of Virginia under the honor system, in which students pledged to obey the code. "From the point of view of the 1930s," he said, "all the right ideas were instilled in me."[67]

"We were all Depression kids," explained one of NASA's top executives. He grew up in an old mill town in southern Massachusetts, where his father delivered milk. "My father worked all the way through the Depression, but we never had much money." Like so many other NASA employees of that generation, the executive took odd jobs when he was young and worked while he was in college. His parents sought to instill in him the values of "hard work, being conscientious and honest, and all those good things. We always had a very open family. I mean, we talked openly about lots of things."[68]

"There were I think fewer opportunities to stray," said a NASA executive who grew up in western Indiana. "There was a recognition in our family that you didn't lie, cheat, steal, or any of those sorts of things. Just the general recognition without being told once a week."[69]

"My mother was a very devout Roman Catholic," said one of NASA's test pilots who moved up to an executive position. His mother found a job and raised the family after his father died. "I would not sit here and pretend that I was perfect. I certainly was mischievous at times during my grade school and high school days. But I think my mother certainly stressed honesty across the board. She was a good person."[70]

"He worked on the railroad," said another NASA executive of his father. "One of the menial tasks they later decided to do with machines when people wouldn't work that cheaply any more." His father, an Italian immigrant, shoveled coal on railroad tenders in upstate New York. "My father was killed in a railroad accident when I was only six months old, so I never really knew my father. My mother, in effect, raised the family, although she did remarry later." When he was fifteen years old, the future executive took a part-time job distributing ice from a delivery truck. "Obviously something in my upbringing made me understand that the only way to break out of this circle that I called the lower end of the economic ladder would be education." His mother could not afford to send him to the university. Scholarships paid his way to college, and he became an engineer and later one of NASA's top rocketeers.[71]

They went off to college, the first generation of engineers, where most of them majored in engineering and some of them majored in science. They emerged from college as members of a professional corps. People who practice a profession, by definition, rely upon specialized bodies of knowledge created through scientific investigation to solve problems. Members of professions are also expected to adopt as their prime purpose "the rendering of a public service" such as medicine or law. People in professions typically resist the intrusion of political considerations into their work, and they are generally hostile to the constraints imposed by bureaucracy. To the professional working inside government, the need to cater to politics and bureaucracy creates criteria that are commonly viewed as irrelevant to the problems that the official is expected to solve.[72]

What causes people to adopt a professional orientation toward their work? For the first generation of NASA engineers and scientists, the professions represented more than a means to earn a salary. Their professional values consisted of more than the education they had received. Most NASA engineers and scientists were first-generation professionals, new entrants to a professional corps. They came from simple backgrounds. They had little opportunity to develop social and political abilities. Their opportunity for advancement rested with their ability to master mathematics and technical skills. NASA recruited people for whom membership in a profession represented a large step up the social and economic ladder. They were committed to professional values not just because they had received an education in engineering or science but because that commitment had a social value as well.

During the first decade of operations, NASA acquired a reputation as one of the most thoroughly professional organizations in the federal government. The social backgrounds of NASA engineers and scientists predisposed them to speak honestly, to work hard at what they viewed as a public service, to discuss space flight as a technical challenge requiring great skill in science and engineering, and to resist the intrusions of politics and bureaucracy as irrelevant to the most important problems they faced.

This is not to say that politics and bureaucracy played no role in the early space program. Politics occasionally did intrude. The location of field centers along the space crescent from Houston to Florida, for example, was dictated as much by congressional politics as by technical considerations. The division of work among centers served bureaucratic necessities as well as technical criteria. NASA officials of the first generation, nonetheless, ferociously defended the importance of placing technical considerations first. "Many NASA managers," observed one, "learned the folly of letting politics and bureaucracy intrude in their service during World War II. If you allowed such an intrusion in war, you were likely to pay with your life."[73] Lives were at stake in the space flight program as well.

Modest beginnings supported the technical orientation of the vast bulk of NASA professional employees. Executive posts in government, however, require the mastery of social and political skills as well as technical ones. The people who rose to the top of the NASA hierarchy were twice as likely as all NASA professional employees of their generation to have grown up in families headed by physicians, engineers, college professors, government executives, and other occupational groups that were at least their professional equals. Growing up in such families seems to have given these people a slight advantage in acquiring the skills necessary to manage people as well as machines. As a whole, however, most NASA executives came from the same modest beginnings that characterized agency professionals as a whole.

The original NASA culture was also imbued with a certain degree of idealism. Professionalism in its classic form requires the bearer to perform a public service, whether it be a doctor curing the sick or an engineer speaking the truth. Two forms of idealism contributed to the NASA culture of the first generation. One was the notion of the space race as the "good war"; the other was the romance of flight.

The airplane was barely twenty-five years old when the first generation of NASA employees was born. Most people traveled by bus or train, if they traveled at all. Flying in airplanes above the ground had a romantic quality that touched many NASA engineers while they were young. Space flight was an even more fantastic idea. One-third of the NASA officials interviewed for this study developed an interest in model airplanes or space flight as they grew up. Others were interested in engines or radios, which propelled them into mechanical or electrical engineering and eventually into NASA.

"Like so many kids," remembered one of NASA's top aeronautical engineers, "you just had that love of watching airplanes and building model airplanes ever since you were little." His father was a truck mechanic who also worked for the Air National Guard. At first, the young man thought that he wanted to be an architect. "The Depression wasn't completely over," he observed. "I came home one day in the middle of summer and there was a gentleman sitting on the front porch talking to my mother. It turned out he was the president of the Spokane Junior College. It was small enough that he was out recruiting students by himself." The young man joined them, and as they sat there talking the college president told him about the civilian pilot training program that the college ran. "And boy, my ears perked up and I thought that sounded good."[74] He signed up, went on to college, and became a NASA test pilot and a flight director.

"I had a love affair with the airplane ever since about 1926," said another one of NASA's top aeronautical engineers. Instead of listening to his teachers at school, he would take aircraft magazines to class and stick them inside his books and read them while his teachers spoke. His parents sent him to a summer camp in Vermont that was devoted to building and flying model airplanes. "I was bound and determined I wanted to design airplanes," he recalled. The director of the camp took him aside. "Look," he remembered the director saying, "you're not going to do any airplane designing if you don't get with it and get a college education and calculus and all the rest."[75]

When the space race began, World War II was a clear memory for most in the first generation of NASA employees. Many of them served in the armed forces. World War II was widely viewed as a moral undertaking, fought to punish Fascist aggression in Europe and Japanese imperialism in the Pacific. Two decades of cold war anxiety followed the Second World War. The Soviet Union took control of Eastern Europe and began to develop thermonuclear arms. Americans heard weekly air raid siren tests and practiced civil defense drills. There was a general sense that a real war could start at any time. The launching of the first two earth-orbiting satellites by the Soviet Union in 1957 came as a tremendous shock to Americans, with their sense of technological superiority. To the people who understood the technology, the second satellite was even more ominous than the first. *Sputnik 1* was a small satellite. *Sputnik 2* weighed 508 kilograms, sixty times as much as the instrument that the United States would launch as *Explorer 1* the following year.

competition w/ USSR

NASA employees of the first generation described the early years of the space race in warlike terms. "There was no question but what the Sputniks and the failure of Vanguard shook this country to its roots," said one NASA executive recalling that year. The United States did not yet have an operational ICBM. Earth satellites and human space flight were as fanciful as expeditions to the stars. The average person, he ventured, had no idea that the Soviet Union could send satellites into space. "That really, as I think back on it, was about as close to Pearl Harbor as this country has seen since then."[76] NASA employees were called upon to wage a major battle in the cold war, and to do so with all of the technical skills at their command.[77] It was not a challenge that readily lent itself to political compromise. The calling reinforced predispositions among the first generation of NASA employees to work hard, to be technically honest, and to engage in public service in a great cause.

It came as quite a shock to NASA professionals when, as the space race came to an end, political considerations began to supercede their professional judgments. This was especially disturbing since the change occurred so quickly after the landing on the Moon. To NASA employees, the lunar landings represented the triumph of their organizational philosophy, a professional accomplishment of the first magnitude. Soon thereafter, the space program was forced to enter the political medium, within which most government agencies exist. NASA professionals had to come back to earth.

"It's important to recognize," said one NASA space flight executive who worked for the agency through its first three decades, "that politicians have very important roles. They have to market and move, they have to get the Congress of the United States and the President behind us." At a certain point, he noted, expressing the creed of the professional, political decisions end and technical decisions begin. "In a similar fashion," he added, "the technician should never try to make the politician's decisions."[78]

Although this dichotomy between political decisions and technical judgments may seem naive to seasoned government watchers, it was pervasively believed by NASA professionals of the first generation. Given the amount of technical discretion allowed agency professionals during the first decade, they actually could separate political influence from engineering criteria on decisions like how to get to the Moon. As NASA entered its second full decade of exploration, considerations other than technical ones began to intrude upon what NASA professionals viewed as their prerogatives. None generated as much concern as the development of the Space Transportation System. Having achieved the first landing on the Moon in the summer of 1969, NASA officials began the search for missions to guide the agency during the next two decades of space exploration. Development of the Space Transportation System emerged as the most likely space flight initiative for the 1970s.

Initially, NASA officials viewed the space shuttle as a vehicle for carrying astronauts and supplies to and from an earth-orbiting space station. They planned to build a fully reusable spacecraft that would replace the traditional blast-off and splash-down approach to human space flight. Such a spacecraft, they estimated, would cost between $10 and $13 billion to develop. White House officials, however, would not approve any new human space flight activity that cost much more than $5 billion. In order to win White House approval for the program in 1972, NASA executives compromised the design of the space shuttle and overstated its cost-effectiveness. Both of these moves contravened the original cultural norms that required engineers to make independent design decisions and technically honest estimates.

To fit the space shuttle into a $5.15 billion development budget, NASA officials abandoned the fully reusable design. They substituted solid rocket boosters for liquid-fueled engines. They promised to make the shuttle cost-effective, which is to say that they promised to operate the Space Transportation System at a significantly lower price per mission than the agency had spent to launch rockets during the first decade. To meet the latter goal, they needed to fly the shuttle frequently—as many as twenty-four times per year. Frequent flights reduced the fixed costs of flying the shuttle by spreading those costs over a greater number of missions. To fly that frequently, however, NASA had to place payloads on the shuttle that would normally be launched by conventional rockets, including commercial and military satellites and the building blocks of the space station to which the shuttle might eventually fly.[79]

"Telling the Congress that we were building a shuttle that was cost-effective when you could calculate on the back of an envelope that it wasn't," said one of NASA's top executives who had by then retired, "I just don't understand why you do a thing like that." He could not understand why the leadership of NASA would perpetuate such a myth. "They can use the back of an envelope like everybody else. You don't have to say it is cost-effective when it isn't. That's a lie. You never lie. Lying is wrong. That always gets you in trouble."[80]

"People thought they were going to do it cheaper in terms of less this, less that, less checks," said another Apollo executive who had helped to launch the giant Saturn V. "They told me they were going to launch this thing in Florida with much less people than I used." That accusation hurt his professional pride. "I didn't think I had wasted a goddamn dime, but they did." Look at the number of people that it actually took to conduct a shuttle launch, he said. A shuttle launch was just as hard as a Saturn V launch. "There are no shortcuts."[81]

How did NASA executives trap themselves into thinking that they could build a cost-effective space transportation system when their experience told them that it was not possible? "I created the argument," said another NASA executive, "so I thought it was great."[82] The President's Science Advisory Committee had told President Richard Nixon that future human expeditions into space could not be justified until the United States greatly reduced the cost of space transportation.[83] "It was clear that one was not going to be able to sustain a major space program without reducing the cost of transportation substantially," said the NASA executive. "There is a simple thing that I learned the hard way, and that is, if you throw away expensive equipment, you cannot reduce the cost of transportation. There's no way that you can make a railroad cost-effective if you throw away the locomotive every time."

"Now unfortunately, the program that evolved as the shuttle was not the program that was going to reduce the cost of transportation." This executive wanted NASA to build the fully reusable shuttle, one that would have cost at least twice as much to develop as the White House wanted to spend. "That was a fundamental error that [the NASA Administrator] made. He should have said, well, if you want a $5 billion program, we'll stay with the Saturn Vs."[84] Faced with budget constraints, NASA closed down the Saturn V rocket program shortly after the 1969 landing on the Moon.

"That was one of the greatest mistakes that NASA made," said a top NASA scientist. The NASA Administrator should have taken the position that the agency could not operate "on a design-to-cost basis." He should have told the President that NASA could not build a space shuttle for $5 billion. "If $5 billion is all the nation can afford, then we can't afford the shuttle. We can't afford to stay in the manned flight business. I am recommending that we close down the manned space flight centers." The program would have been delayed. The Administrator probably would have been fired or forced to resign. In the long run, he predicted, "Congress and the administration would have found the funds."[85]

In spite of the political compromises that characterized the second decade of space flight, NASA officials continued to recruit people who fit the traditional mold. They continued to recruit people for whom engineering or science represented a big step up the social ladder. A majority of the professional employees recruited by the agency between 1970 and 1988 still had to work their way through college (see the appendix: origins). The career executives who led the agency during the second and third decades of space flight espoused the same middle-class values as the first generation of NASA scientists and engineers. At the same time, the circumstances within which the agency had to operate changed. No longer could agency leaders rely upon a space race to promote the supremacy of technical judgment. No longer was the space program viewed as a good war or a moral crusade. Public-spirited idealism shifted to other issues, like the environment and the movement to end the fighting in Vietnam.

NASA officials clung to the old values. Many were deeply disturbed by the changes they saw. Few of the people interviewed for this study blamed NASA's decline on the changing of the generations in American society, however. Instead, they tended to blame the top executives at NASA for not protecting the professional character of the organization. The biggest change in NASA from the 1960s to the 1980s, said one of NASA's top project managers, was "just a change in the character of the overall strength at headquarters. . . . The administrators were very strong, very protective of the organization," he said, referring to the first generation of leaders. He recalled an incident during the visit of President Kennedy to the Marshall Space Flight Center in the fall of 1962. Wernher von Braun, who was then the center director, Brainerd Holmes, NASA's Chief of Manned Space Flight, and Jerome Wiesner, the President's Science Advisor, got into a heated discussion on the best way to land on the Moon. Wiesner was skeptical of NASA's decision to attempt a lunar orbit rendezvous. "It was extremely interesting," the project manager recalled. "They just stood there arguing, and President Kennedy and Vice President Johnson were listening. And, finally, President Kennedy decided he'd had enough, and so they walked away."

The NASA people, along with some of the President's aides, got on a bus to go to the next stop. "David Bell, who was Director of the Budget, was there, and the Secretary of the Air Force was there and some high-ranking professional people and other people. And old Jim Webb [the NASA Administrator], he stood up at the front of the bus in the well. He said he wasn't going to have a bunch of [outsiders] tell him how to run his program. They [the NASA engineers] had worked the thing and he just wasn't going to turn around. He just said, 'Hey, we're running this show and we're going to run it.'" Webb overruled the President's Science Advisor. The NASA Administrator, the project manager concluded, "had a lot to do, as much as any, I guess, with setting the tone for NASA."[86]

NASA officials worried that the intrusion of politics and bureaucracy would compromise the performance of the agency. "This is a very unforgiving business," said one space flight executive, "and you can't afford to be wrong." He went back to what George Low, one of NASA's top career executives, used to say: "It's attention to detail and being right all of the time. . . . Each of the technical issues was aired and discussed in great detail, and there were differing opinions. There was a lot of emphasis put on testing and being sure that you were really ready to fly, that you had as much confidence as you could that the system was going to work the right way when you went to fly." These things, the executive believed, began to change in the 1970s, when budgets became more of a problem. The people who moved into top agency positions did not have the same ability to protect the old NASA values. "The character of the agency," he concluded, "really started to change."[87]

4 ▲ Becoming Conventional

*Bureaus, like men, change in predictable ways
as they grow older.*—Anthony Downs, 1966

Many aspects of NASA's technical culture were well developed at the time that the new space and aeronautics agency started work in 1958. The emphasis on testing, the commitment to in-house technical capability, and the strong belief in hands-on activity were deeply rooted in the predecessor organizations. Forty years of aeronautical research supported the technical culture of the NACA; the German rocketeers and the ABMA had been working together for twenty. Values added during NASA's formative years easily fit the old traditions—the attitudes toward risk and failure, the frontier mentality, and the belief that the organization attracted exceptional people. All of these values built logically upon elements of the NACA and ABMA cultures. NASA's technical culture flowed from a well-established tradition.

People in the predecessor organizations adopted the technical culture because it worked well for the tasks confronting them—aircraft research and rocket development. The technical culture was also well suited for small NASA projects, such as research activities and satellite development, that involved only a few centers and a limited number of contractors. The people who ran the NACA and the ABMA, however, had never managed a macroengineering project as large as the Apollo expedition to the Moon. They had never been responsible for a project that involved so many contractors and participating centers. "We were faced with an entirely different set of goals and an entirely different environment," said one of the people who helped organize NASA's first steps toward the Moon.[1]

Just one month before NASA came into being in the fall of 1958, President Dwight Eisenhower approved what became known as Project

Mercury, the program to put America's first astronauts into space.[2] The NACA/NASA employees who organized Project Mercury quickly recognized the special nature of the challenge created by the new space flight program in an organization heretofore devoted to flight research. "We now had to build something. We now had to fly something that we built. We now had to interface with the contractors to get that done. We had to build an organization. We had to make things happen that we had not ever been associated with before."

"Particularly in the early days of the Space Task Group [the NASA group that ran the Mercury program], we were faced with having to do not one job but ten. . . . We had to write specifications. We had to develop an RFP [a request for proposal, specifications from which contractors make their bids]. We had to develop an organization to carry it out. We had to do detailed testing of hardware, all of which we had experience with, but had not had to manage either from a budgetary sense, a contractual sense . . . with other people that had to do it on our direction."[3]

Two and one-half years later President Kennedy approved Project Apollo, the crash program to reach the Moon. The lunar goal dwarfed the requirements created by Project Mercury and overwhelmed the technical culture rooted in the NACA and the ABMA. NASA officials struggled with the management problems of this large engineering program. These problems were especially troublesome in Project Gemini. The Gemini space flight program was designed as an intermediate link between the short-duration, single-person space flights of Project Mercury and the lunar expeditions of Project Apollo. Approved in late 1961, Project Gemini called for twelve flights during the mid-1960s to test the capability of humans and machines to function in space over increasingly longer periods of time and practice the rendezvous techniques that would be necessary to fly to the Moon. By 1963, the Gemini space flight program was over budget and behind schedule. Technical problems plagued the Titan II launch vehicle and the capsule landing system, a parachute that was supposed to work like a glider.[4] To put the space flight program back on track, NASA executives reorganized their manned space flight program.[5] Although the problems had erupted in Project Gemini, the changes had their greatest effect on Project Apollo.

Organizing for Apollo

To put the human space flight program back on track, NASA executives took the technical culture rooted in their field centers and imposed over it a centralized management system. That management system had its roots in the Air Force Ballistic Missile Program. "What made NASA reasonably good in the 1960s," said one of the people brought in to change it, "was really a blending of three cultures. It was a blending of the German rocket culture of von Braun and company, the NACA experimental aircraft background, but then we had a reasonable infusion of some very key people from the Air Force Ballistic Missile Program. . . . It was really bringing those three cultures together that made the thing work."[6]

As part of a 1963 reorganization, NASA Administrator James Webb changed the leadership of NASA's manned space flight program. He brought in George Mueller (pronounced "Miller") as the Associate Administrator for Manned Space Flight. Before coming to NASA, Mueller had worked on Air Force missile programs as a vice president at Space Technology Laboratories, a California aerospace firm. Mueller recruited Air Force General Samuel C. Phillips to take charge of the overall management of the Apollo program. Phillips had been the Vice Commander of the Air Force Ballistic Systems Division, and had a clear understanding of the management techniques employed in the Air Force Ballistic Missile Program. To take charge of the development of the Apollo spacecraft, Mueller dispatched Joseph Shea to the Manned Spacecraft Center at Houston. Shea had joined NASA in 1962 as part of an earlier reorganization that had brought in other ballistic missile experts.[7] Prior to arriving at NASA, Shea had worked as a missile systems engineer on the Air Force Intercontinental Ballistic Missile Program. Mueller dispatched another Air Force general, Edmund F. O'Connor, to run the Industrial Operations Division at the Marshall Space Flight Center, which housed the program offices for the large rocket program. Others joined the migration from the Air Force missile program to the civilian space agency. NASA's success in achieving the goals of the Apollo program was due in large measure to the tension between the Air Force approach to program management and NASA's traditional technical culture. The system worked remarkably well during NASA's formative years.

The people with Air Force backgrounds brought in to manage the manned space flight program did not have a high regard for NASA's ability to manage large programs. "NASA had considerable technical depth," said one of the Air Force people, "but almost no program management experience."[8] Nobody knew how to do program management or work with industry on large programs. Another top NASA executive agreed. For a research organization, he offered, NASA's management capabilities were excellent. As for NASA's capability to run large programs, "I wouldn't say they had any." The people at the Marshall Space Flight Center "had a very well-developed management capability, but again, principally for doing internal work." Technically, they were capable. But their management capability "was oriented toward managing their own resources and their own developments, just as were the NACA folks."

This point of view seemed to contradict one of the tenets of NASA's technical culture. Many NACA employees believed that in-house research provided a training ground for project managers. "If you're thinking of a project like a guidance system, yes," said the executive. In-house work had provided management training for small projects. "If you're thinking about a program," he added, then that experience "didn't reside within any of the Langley folks." In neither case, the executive said, were the NACA or the ABMA people accustomed to dealing with industry on large projects. "They did their own work. They had their own technicians. They built their own things."[9]

The people with Air Force backgrounds sought to make a number of important changes in NASA practices, changes that were in each case contrary to the tenets of the old technical culture. First, they wanted the NASA people to place more trust in the capabilities of industry. "There had grown up a situation on the Mercury and Gemini programs where the operations people who were responsible for the launch at the Cape didn't trust the quality of the hardware coming out of the factory," said one of the Apollo program managers. "They had actually fallen into a situation where they would take the spacecraft before it was completely checked out and bring it down to the Cape and almost take it apart and put it back together again."[10] This practice arose from NASA's tradition of in-house work and hands-on activity, which had produced a desire among NASA employees to penetrate the work of contractors.

"They knew how to manage them," one official said of the German rocketeers and their approach toward contractors, "by taking them over." The rocket team at the Marshall Space Flight Center exercised close supervision over their contractors. "Management was exercised through very deep penetration by civil servants into every technical detail and every design decision." This was a tradition that the rocket team had formed in Germany, the official added. European governments allowed their civil servants to engage in practices that could be viewed in the United States as interference in industrial affairs. For the rocket team, "the continued breadth and depth of civil servant involvement became a trademark, a source of pride, and it created within the center a considerable feeling of independence."[11] The old employees naturally wanted to continue that approach.

Officers working on the Air Force Ballistic Missile Program developed a different tradition. They relied extensively on contractors but did not converge with them to the same degree. In their ballistic missile program, for example, Air Force officials hired a private firm to help coordinate the program rather than conduct the systems engineering function in-house.[12] When the Air Force people attempted to impose this approach on NASA, they met resistance. It was quite a point of contention. "One of the things that Mueller and I wanted," said one of the people from the Air Force program, "was to force the discipline back into the factory and use the same check-out equipment in the factory as at the pad. That was a big issue at the time I was there."[13]

To verify contractor work, agents of the traditional culture within NASA sought to conduct extensive flight tests. This kept faith with the old traditions of verification and in-house work. Marshall Center employees, responsible for the development of the Saturn rockets that would propel the astronauts toward the Moon, wanted to conduct those tests incrementally. They wanted to test the first stage of the Saturn rocket, improve and verify its performance, test the second stage, which could be done simultaneously, separately test the third stage, and so on.

The Air Force ballistic missile philosophy, on the other hand, placed much less emphasis upon flight tests. They relied more upon ground tests. By conducting ground tests of ever increasing complexity, they reduced the need for extensive flight testing. One of the executives from the Air Force program explained their reluctance to adopt NASA's approach. "At the time," he said, "reliability engineers argued that man-rating a vehicle like the Saturn V should require some element of statistical confidence. An oft-cited goal was 90 percent confidence of 95 percent reliability." Even those statistics would produce a nearly 50 percent chance of losing a Saturn V during the eleven Apollo launches with astronauts on board. Moreover, to verify those numbers, NASA would have to conduct a large number of Saturn V test flights. The executive calculated that it would require forty-five flight tests. "The decade would be long ended before that could happen."[14]

In what was clearly the most controversial move made by the Air Force group, George Mueller announced in the fall of 1963 that NASA would drop the traditional step-by-step flight tests of rocket and space-craft components. Instead, he told NASA engineers to assemble all three stages of the Saturn V rocket along with the command and service module—as if they were ready to fly to the Moon—and conduct just two or three unmanned test flights of the whole system.

Von Braun's rocket team was horrified when it heard of the concept. It was totally contrary to their rules of test and verification. "We had never done anything like that. We always had to build up step by step."[15] If NASA lost one of the lower stages of the rocket during an all-up test flight, von Braun's team argued, they would learn nothing about the performance of the upper stages, nor the problems of bringing the command module back to earth at translunar speeds. Those parts of the test would be thrown away. NASA experienced plenty of problems during ground tests of the tricky hydrogen-oxygen-fueled S-II second stage—faulty instrumentation, an engine fire, and a rupture of the liquid hydrogen tank—so a failure on a flight test was more than a remote possibility. "We didn't like this at all," said one of von Braun's associates, "because you know by your experience nothing works from the beginning completely. There are always a few things which you want to iron out. And therefore, if you go with an all-up concept, that is far too dangerous."[16]

The Apollo executives insisted on testing everything at once. "The alternative [sequential flight tests] really didn't gain you anything," one argued. "You didn't decrease the risks by testing sequentially; you only spread the risks out." This executive was willing to accept the risk that NASA might lose one or two rocket stages. "But in that case, you were going to be grounded for a while anyhow, so you hadn't lost anything, other than the hardware." Without all-up testing, NASA had only a slim chance of achieving President Kennedy's goal of reaching the Moon by the end of the decade. "The risk was there in any event," he concluded, and by moving to all-up testing their progress certainly could be "a lot faster than it otherwise would possibly be."[17]

Mueller insisted on his approach; the German rocket team insisted on its. Mueller's position prevailed. He outranked the rocket team, and the demands of NASA's schedule outweighed the incremental testing philosophy. Mueller's position, it should be noted, was not contrary to all of NASA's technical norms. Risk was normal, and failure (on an unmanned test flight) was allowed. "I still admire him for that courage he had," said one of the German rocketeers twenty years later.[18]

Mueller's gamble paid off when, on November 9, 1967, NASA officials successfully fired all three stages of the Saturn V rocket along with the command and service module—without astronauts on board—through a full earth orbital test flight. The flight could have just as easily failed, as the next all-up test flight did. The performance of the rocket during the second all up test on April 4, 1968, was so poor that NASA officials rated the mission unsuccessful. That was all the Saturn V tests that Mueller needed, however. Eight months later, American astronauts were circling the Moon.

The diminished number of flight tests did not mean that NASA reduced testing altogether. Testing was too much a part of the technical culture to do that. Instead, the Air Force ballistic missile people sought to test on the ground what the bearers of the old technical norms previously tested in flight. All-up testing required "a most comprehensive ground test program, which prevented the revelation of hardware weaknesses in flight."[19] That tended to push work back toward the contractors, who were in a better position to test components on the ground than in the air. Pushing work toward contractors was a key objective of the Air Force people.

Officials from the Air Force Ballistic Missile Program also worked to strengthen program management within NASA, especially at NASA headquarters. NASA officials had already established project management offices at their field centers to oversee spacecraft and rocket development. Air Force officials wanted to establish overall program management offices for each of the major space flight programs at Washington, D.C., and they wanted all of the offices both at headquarters and in the field to perform more management activities. In the ballistic missile program, Air Force officials had worked out a highly disciplined management system. The system allowed them to formulate requirements, manage costs, control schedules, check performance, and plan operations. "It had to start with the technical specifications and flow on through the disciplines of design reviews . . . configuration management . . . responsibility accounting . . . design certification reviews . . . final readiness reviews," one Air Force officer explained. "That was a pattern I think of under system engineering and technical controls. There are a bunch of subitems in it like weight budgeting and control, thermal budgeting and control, electrical power, a number of those things. Another but parallel aspect, of course, is program planning and control, which was not a discipline that NASA had developed either."[20]

All of these strange sounding requirements had one unmistakable effect. They centralized the management of the Apollo space flight program to a degree far beyond that which the bearers of the old technical culture had allowed. The NACA and the ABMA were decentralized organizations in which engineers and scientists carried out the work of the agency in well-insulated centers located far out in the field. Both NACA and ABMA officials relied upon engineering divisions or laboratories at the various centers to carry out technical work. Technical decisions were often reached in executive sessions with the laboratory chiefs and the center director sitting together. Once the large rocket and space flight programs got under way, the center directors set up project management offices to monitor the work of contractors and keep the programs on track. Project managers in the field shared power with people in the engineering divisions and technical labs. By emphasizing program management, the Air Force officials enhanced the power of the project managers.[21]

For a program as complicated as Project Apollo, the centralized approach had to prevail. In Washington, NASA higher-ups appointed an overall Apollo Program Director and put him on the third floor of the headquarters building. From 1964 to 1969 this post was held by Air Force General Samuel Phillips. To make sure that he could manage the Apollo expedition, NASA officials reorganized the agency so that the three field center directors with primary responsibility for the expedition (the Huntsville, Houston, and Florida center directors) reported directly to the headquarters office in which the Apollo program director worked. NASA executives then allowed the Apollo program director in Washington to hire his own cadre of contractors to perform overall systems engineering and integration work and monitor the performance of other contractors, a practice taken directly from the Air Force Ballistic Missile Program. In this way, the Apollo program director gained the authority to coordinate the overall expedition and not be dependent upon the field centers for this work.

Throughout the 1960s, the old technical culture and the new forces of organization coexisted in a state of more or less healthy tension. To the extent that an Apollo culture can be said to exist, this was it. Engineers and scientists representing the traditional culture brought their technical values to the table. Those values elevated technical criteria as the primary means for making space flight decisions and depressed the importance of bureaucracy and politics. The program managers (often engineers themselves) imposed official procedures and schedules that kept the expedition on track.

People working within this overall culture frequently disagreed, as the controversy over all-up testing reveals. While the outcome of such disputes was important, so was the fact that different parties felt free to present their points of view in an open way. Conflict between different points of view was a normal way of doing business in the early NASA organization. The willingness to discuss such conflicts openly was part of NASA's original culture.

"There was a very strong teamwork culture that developed," said one leading Apollo program official, "which was wide open communications." People fought out their differences. Before contractors and civil servants at different centers learned how to do this, the Apollo official observed, they tended to "hold the revelation of a problem until they had it solved."[22] People at NASA headquarters set up a number of formal mechanisms to discourage this: special management councils, special review teams, and an Apollo executive committee. Some teams brought people together to discuss project problems; other teams went into the field to solve them. One team, for example, worked with the lunar module contractor when costs for that system soared. Managers at NASA headquarters set up formal systems for reviewing the design of spacecraft and rocket components. They set up formal mechanisms for checking flight readiness before committing to a launch. Underlying these formal mechanisms was the cultural norm of full and open communication. Higher-ups liked to portray this as a norm of teamwork, but people in the trenches knew a different reality. Open communication meant a lot of conflict. "This culture," explained a headquarters official working to develop the NASA program management system, "if people tell you everybody was lovey-dovey through this business, that is not true. We had lots of fights among key people."[23]

The presence of strong technical norms encouraged a certain amount of arrogance, or what people who observed these battles viewed as arrogance. Engineers and scientists who were reasonably confident of their technical experience, placed in meetings where they were free to express their own points of view, tended to treat opposing judgments as rooted in ignorance. The natural rivalry produced by strong personalities with different responsibilities added to that. Discussions could get quite heated. There was not much of a tradition of holding back.

NASA's organizational culture reached its apex during the preparation for the Apollo flights to the Moon. Technical discretion and organizational skill balanced each other. That balance proved to be ephemeral. As NASA matured, the technical culture grew weaker. So did the centralizing forces brought in to manage Project Apollo. In their place a more conventional form of government organization arose. NASA grew bureaucratic. Maintaining NASA's organizational culture as practiced by the first generation of employees turned out to be most difficult to do.

Aging and Organizational Change

As any newly created government agency matures, it encounters forces that challenge its original culture. Government-wide pressures to impose more rules and regulations increase the agency's bureaucratic burden. Cycles of expansion and contraction modify overall operations. The type of people who are attracted to the agency may change. Singly or in concert, these forces affect the culture of an agency in a number of ways. The agency may lose that flexibility it enjoyed during its formative years. It may become more bureaucratic. Its managers may grow more conservative or averse to risk. Its officials may become more preoccupied with the survival of the organization. Officials in agencies so affected find their organizations propelled through a life cycle that transforms the character of the original institution. So powerful are these tendencies in government that some scholars find organizational life cycles to be as inevitable as the processes of birth, maturity, and decay in living organisms.[24]

Officials at the National Aeronautics and Space Administration did not escape these trends. Forces of change began to affect internal operations before the new space agency turned ten years old. The effects of contracting resources and diminished public support, coming as they did after a period of rapid expansion, were especially dramatic.

Scholars find a predictable pattern in the life cycle of government bureaus. In the beginning, sometimes prior to their official creation as independent agencies, new organizations go through a period of vulnerability. Their jurisdiction is not fully established; their relationships with potential clientele not settled. The U.S. Army Air Force, for example, struggled to maintain its newfound responsibilities in the period following World War I. Not until its rapid expansion during the second world war did the Air Force emerge as a coequal branch among the military services. Shortly after NASA's creation, its leadership had to repulse an effort by Air Force officials bent on taking control of the national space effort.[25]

Agencies that survive infancy generally go through a period of rapid growth. Their work force expands as the agencies gear up to perform functions assigned to them. Their budgets grow. Opportunities for promotion increase. Public support for their activities remains fairly high, a consequence of the urgency associated with their creation and a brief honeymoon period. Following its creation in 1970, for example, the U.S. Environmental Protection Agency embarked upon a ten-year period of budgetary and personnel growth. New agencies that expand in this fashion become permanent fixtures on the government scene. It is quite hard to kill off a government agency once it emerges from its initial period of expansion.[26]

As an agency expands, members of the organization encounter fewer obstacles to the maintenance of a nonbureaucratic culture than at practically any other time in the life of the agency. The fruits of expansion make this so. Opportunities for promotion and the allure of agency operations attract the "right sort" of people to the organization—employees who tend to be young, flexible, and willing to learn. The social importance of the agency's functions provides a rationale for concentrating resources on the work to be done instead of spending time protecting turf. Rendered relatively secure by an expanding work force and budget, agency executives avoid many of the internal battles that preoccupy officials in less fortunate organizations. In order to help the agency get started, politicians may waive many of the formal administrative requirements that bog down older, more established bureaucracies. In setting up the Tennessee Valley Authority (TVA) in 1933, for example, the U.S. Congress exempted the new agency from burdensome administrative regulations during its formative years and placed its governing board in Knoxville, Tennessee, in part to distance it from the constant oversight imposed on federal executives housed in Washington, D.C. After an initial period of political insecurity, the TVA emerged as a major agent of economic development in the seven-state region.[27] The period of expansion is generally a golden time for the members of a new agency and their cause.

Inevitably, the period of expansion must end. Many factors contribute to its cessation. Some agencies attain the size appropriate to their functions. Others complete their initial tasks. For many, political support wanes as new issues force their way onto the public agenda. Leaders of other agencies with which the new organization must compete for resources resist further growth. Within the government as a whole, the work of the agency acquires less priority or less visibility. The agency budget (in constant dollars) or its work force commonly contracts. The initial growth of the U.S. Peace Corps, for example, which was established in 1961, was followed by a ten-year period of retrenchment which saw its budget (in constant dollars) and volunteer work force cut in half.[28]

For whatever reasons, agency leaders who had once enjoyed the advantages of growth must at this point grapple with the problems imposed by the lack of it. The end of expansion poses a severe challenge for agency leaders attempting to maintain the culture built up during the period of growth. In order to flourish, such institutions must substitute qualitative growth for quantitative growth. This is a demanding requirement even under the best of conditions. Only a few institutions, such as well-respected universities, have been able to follow this strategy, relying upon their prestige to attract innovative people even as their budgets and work force stabilize.[29] An organization for which both qualitative and quantitative growth has ended, by contrast, is ripe for cultural transformation.

The history of the U.S. civilian space program provides a textbook example of the expansion-contraction cycle. NASA entered the world on October 1, 1958, with some 8,000 employees on board. Eighteen months later, approximately 4,000 employees transferred in from the Army Ballistic Missile Agency. In spite of this rather large birth weight, NASA did not receive its overall mission until the spring of 1961. For two and one-half years, the only manned space flight mission for which the agency received authorization was the relatively small Project Mercury, intended to sustain a single astronaut in low earth orbit for about one day. The long-range direction of the NASA space flight program remained uncertain until 1961, when President Kennedy decided to dispatch Americans to the Moon.[30] Kennedy created the macroengineering orientation for which NASA became famous.

The lunar decision established NASA's niche in the executive branch and—at least temporarily—silenced critics who had argued for a space program based more on science and less on large-scale engineering endeavors. Its period of infancy at an end, NASA expanded rapidly. The number of NASA civil servants more than doubled from 16,000 at the end of 1960 to 36,000 employees in 1966. Its budget expanded more than eightfold. By the 1965 fiscal year, NASA officials were spending over $5 billion per year: 0.8 percent of the entire U.S. gross national product or 4.4 percent of all federal outlays that year. NASA's organizational culture blossomed. The technical orientation brought in from the predecessor agencies merged with the organizational requirements imposed by large-scale engineering projects to create the culture of the first generation.

The period of expansion was followed by a period of retrenchment. Even as the United States prepared for the actual flight to the Moon, NASA spending and hiring fell. In constant dollars (adjusted for infla-

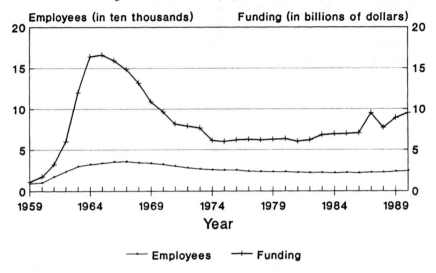

Figure 3   NASA Employees and Funding

Employees (in ten thousands)   Funding (in billions of dollars)

Year

———• Employees   ——+— Funding

Source: NASA, *Pocket Statistics*, 1992.
Note: NASA employees include both permanent and temporary personnel on board at the end of the fiscal year. NASA funding (total appropriations by fiscal year) is calculated in constant 1982–84 dollars using the consumer price index.

tion), the NASA budget began to contract in 1967 and continued to do so until 1975. Its budget as a percentage of the gross national product followed a similar rise and fall. By 1975, the United States was spending just 0.2 percent of its gross national product on NASA—one-quarter of the spending rate during the peak Apollo years. NASA civil service employment peaked at 36,000 employees in 1967 and fell to 22,000 by 1982. The number of contractors working for NASA was similarly reduced from over 300,000 in 1966 to just 100,000 by 1972.[31]

Public support for the space program likewise declined. It is worth noting that public support for extensive space spending was never especially high, even during the formative years. When asked in May 1961 whether the United States should spend up to $40 billion "to send a man to the moon," respondents to a Gallup public opinion poll responded 58 to 33 percent against the initiative.[32] President Kennedy decided to go anyway. In 1965, the proportion of people who wanted to see space spending cut outnumbered the proportion who wanted to see spending increased by a ratio of two to one. As NASA continued preparations for the lunar expedition, public support grew even weaker. It weakened as the United States prepared to go to the Moon, it weak-

Table 2   Public Support for Space Exploration

| | Percentage of Respondents | | | |
Year	Spending Is Too Little; Spend More	Spending Is Too Much; Spend Less	Spending Is About Right; Stay the Same	No Opinion
1965	16	33	42	9
1969	14	40	41	5
1973	7	59	29	5
1980	18	39	34	9
1982	12	40	41	7
1987	33	27	34	6
1988	30	18	48	5

Sources: George H. Gallup, *The Gallup Poll, Public Opinion, 1935–1971* (New York: Random House, 1972); Elizabeth H. Hastings and Philip K. Hastings, eds., *Index to International Public Opinion, 1979–1980* (Westport, Conn.: Greenwood, 1981), and *Index to International Public Opinion, 1982–1983* (Westport, Conn.: Greenwood, 1984); "The Public's Agenda," *Time,* March 30, 1987; *The U.S. Space Program,* Media General/Associated Press Public Opinion Poll 21, June 22–July 2, 1988.
Note: The question in 1965, 1969, 1987, and 1988 asked whether the respondents would like to see the amount of money spent on space exploration or research increased, decreased, or kept about the same. In 1973, 1980, and 1982 the wording was changed to elicit a response to whether the amount of money was too little, too much, or about right.

ened as the United States went to the Moon, and it weakened when the United States stopped going to the Moon.[33] Public support rallied after the fire that took the lives of three astronauts and it rose when the first U.S. astronauts circled the Moon—but the long-term trend was clearly one of decline. By 1973, with the lunar expeditions at an end, the proportion of people who wanted to see space spending cut outnumbered the ones who wanted to see it increased by a ratio of eight to one.

Much of the public support for an aggressive space program in the early years was motivated by a desire to carry out missions superior to those of the Soviet Union. By the time that the United States landed on the Moon, that motivation had largely disappeared. Other priorities had moved forward. A 1973 opinion poll ranked space exploration next to last on a list of eleven government priorities. The space program ranked well below health, the environment, the problems of the cities, crime, drug addiction, and the nation's education system. Even welfare spending was more popular that year. The only program less popular than space was spending on foreign aid.[34]

Where NASA officials had once reaped the benefits of expansion, they now had to grapple with the effects of contraction. Some of the

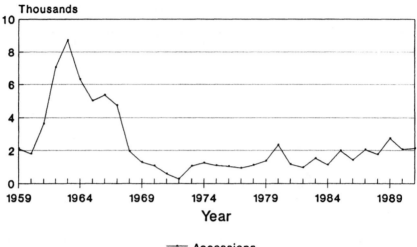

Figure 4 NASA Hiring

--- Accessions

Sources: Jane Van Nimmen and Leonard C. Bruno, *NASA Historical Data Book,* vol. 1, *NASA Resources, 1958–1968,* SP-4012 (Washington: NASA, 1988); and NASA Office of Human Resources and Education, "The Civil Service Work Force," annual (Washington).
Note: Accessions of permanent employees by fiscal year.

effects proved quite painful. With budgets and payrolls shrinking, the agency had to cut back its hiring of new employees. During the formative years (1959 through 1967), NASA officials brought in an average of 4,969 new permanent employees each year. So rapid was the infusion of new blood that in both 1962 and 1963 30 percent of NASA's employees were that year's new hires. In late 1967, nearly two years before the landing on the Moon, NASA hiring fell to pre-Apollo levels.[35] In 1972, it fell to just 264 new permanent employees. In 1972, 1973, 1974, and 1975, agency officials dismissed agency workers through a device known as reductions in force. The lay-offs had a devastating effect on agency morale.

With the flow of new blood constricted, NASA began to show signs of age. The work force grew older. The number of young people entering the work force declined. Engineers and scientists, the core professional group within NASA, advanced in years. In 1966, the year before total agency employment peaked at nearly 36,000 people, 37 percent of NASA's scientists and engineers were between twenty-five and thirty-four years of age. The average age of all NASA scientists and engineers stood at slightly more than thirty-eight years. By 1982, the average age of all NASA scientists and engineers had risen into the

Table 3   The Aging of NASA

Year	Percentage of Scientists and Engineers 25 to 34 Years of Age	Average Age of Scientists and Engineers
1966	37	38.2
1970	32	39.0
1974	20	41.6
1978	12	43.8
1982	13	44.5
1986	20	43.8
1990	31	42.0

Sources: NASA, "Age Distribution of Permanent Employees (1966)"; NASA Office of Human Resources and Education, "The Civil Service Work Force," annual (Washington).

Figure 5   NASA Promotion Rates

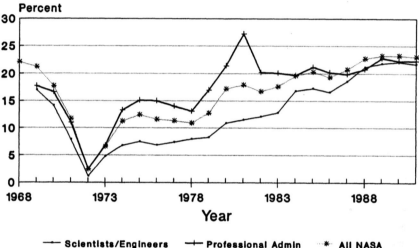

— Scientists/Engineers   —— Professional Admin   ···*·· All NASA

Source: NASA Office of Human Resources and Education, "The Civil Service Work Force," annual (Washington).
Note: Percentage of employees in that group promoted that fiscal year.

mid-forties. The proportion of scientists and engineers aged twenty-five to thirty-four fell to just 13 percent. Not until the original cohort of NASA employees reached retirement age in the 1980s did the average age stop growing.

As NASA's work force grew smaller and the proportion of senior employees grew larger, opportunities for promotions naturally declined. In 1968, the year before NASA landed the first astronauts on the Moon, the chances for any NASA employee to win a promotion were better than one in five. In other words, more than 20 percent of all NASA employees received a promotion of some sort that year. The same odds continued in 1969. After the lunar landing, they fell dramatically. Only one out of forty, or about 2.5 percent of all NASA employees, received a promotion in 1972. Promotion opportunities did not fully rebound until the following decade.

Social scientists have speculated about the cumulative effect of changes such as these on the machinery of public administration. Four effects have special importance: the loss of flexibility that comes with age, the tendency of agencies to grow more bureaucratic with time, the likelihood that their executives will become risk-averse as they and their agencies age, and the inclination of agencies struggling to survive to alter their methods of operation. The noted effects, of course, are not confined to agencies in their initial cycle of growth. They may affect mature institutions as well, especially those emerging from later cycles of expansion and contraction. Young agencies at the end of their period of expansion, however, seem especially vulnerable to these effects.

Decreasing Flexibility

In the beginning, said a NASA executive from the first decade, "the whole administrative system was very flexible." The evidence supports this impression. The conditions under which NASA functioned during the early years—what social scientists like to call the agency's environment—allowed NASA employees to exercise a great deal of flexibility in the administration of the infant space program.

One of the most prevalent forces affecting flexibility in government operations is the civil service system. Civil service rules can severely restrict the ability of public executives to hire and fire the right personnel. The first NASA managers, however, experienced little difficulty fitting new recruits into the civil service system. NASA's personnel officers developed a special technique for working around civil service classification procedures. It was, said the executive, "the most beautiful way to classify engineering jobs I have ever seen." NASA personnel officers produced what looked like a Sears, Roebuck catalog of jobs. A new recruit would look through the catalog for a category that resembled the work that he or she planned to do. "The Civil Service Commission finally clobbered us over that kind of stuff," said the executive. "They tried to take all of our flexibility away." The civil service minions insisted that NASA follow standard position classification procedures.

"We would never have been able to do what we did during the early years," the executive announced, "under the current kind of civil service bull—— that goes on in this town. Not that we broke any laws. But we used our flexibility to the nth degree."[36]

NASA was helped by the availability of positions that fell outside of normal civil service restrictions. "Those were jobs where you could offer a person a position much like you would in industry," explained one of NASA's top scientists. "You could say in those days, I will give you $15,000 or $16,000 a year and you come to work, and the person would come to work the next week. You didn't have to go through the civil service regulations. You weren't limited by the civil service grade structure and all that."[37] NASA's enabling legislation authorized the Administrator to pay up to $19,000 per year for two hundred and sixty scientists, engineers, or administrators. The number of special pay positions nearly tripled as NASA grew.[38]

Rapid growth and increasing budgets during the formative years also lent flexibility to the infant space program. "The other thing that you could do in the early days," said the space scientist, "was that you could take a chance on a person, and if he didn't work out or she didn't work out, you could kind of shove them off to the side and you could bring somebody else in and you would rename the job or something like that. You could move along."

NASA's ability to push aside nonperformers declined when the agency stopped growing. "That became impossible," the space scientist recalled. "I can remember in the late sixties making a conscious, deliberate decision to keep an incompetent person on because he was fighting the dismissal." The scientist realized that he would have to assign one of his good people to build the case. "I would have to put one full-time person to work, just to supervise him, to give him clear, accurate instructions, in writing; and then prepare the clear, accurate, verifiable documentation that would show that he had not performed. And I said to myself, it is better to simply keep him there than it is to divert a good person to firing him."[39]

NASA's early flexibility extended beyond personnel matters to the process of starting up new activities. One of the agency's first executives recalled the process by which NASA found the land to set up the Goddard Space Flight Center. "We had a pretty free hand," he mused. As Congress debated NASA's creation, future NASA executives looked for a spot to house the new center. The search became more urgent as top officials discussed the transfer of the Vanguard satellite group to NASA. "It was obvious that they needed a permanent home," the executive explained. "I went to [Deputy Administrator Hugh] Dryden and I said, 'Hugh, we've got to find some place for these people. We've got to build a lab for them.' And Hugh said, 'Well, that's interesting, because last night I was at a meeting with one of the fellows from the Agriculture Department and he came up to me and asked me if we needed to set up any facilities in the Washington area, and I told him I didn't think so. But, he said, maybe I ought to call him and see what I can do.'"

NASA needed about five hundred acres for the new center. The Agriculture Department had five hundred acres of surplus land on one of its research centers in Greenbelt, Maryland, just outside Washington. "So we went out and looked at their place and I picked out the Goddard site." The executives put in a request for funds to start work on the center. "When it came to be heard at the budget hearings, the congressman who was in charge of us, our Texas friend, he said. 'I don't know about this.'" Representative Albert Thomas of Texas, who headed the Appropriations Committee through which all NACA and NASA funds passed, apparently had not been informed. Thomas wanted NASA to locate a NASA field center in his home state. "We said, well, we need a lab. We've got these people. We've got to put them someplace and if we don't put them in Washington they will leave the program." The Appropriations Committee balked. Its members cut NASA's overall construction budget nearly in half and suggested that the agency defer work on the Goddard site. Agency executives went ahead anyway and allocated $3.9 million to start constructing facilities. "That was how easy it was," the executive recalled. "If you tried to set up a new lab now you would have every congressman in the country looking for the space. . . . We could do things ourselves, you know, reach solutions without consulting with outside committees or anyone else. Just use good judgment and move."[40]

Looking back, one of NASA's science directors remembered how much easier it was to get projects approved in the early days of the space program. "I didn't have to justify anything to anybody. I didn't have to make presentations or put on dog-and-pony shows for visiting committees who were going to then decide whether or not the project was appropriate or anything like that. We just did it." The science director worked on the staff of the Goddard Space Flight Center. The scientists and engineers who worked with him were too busy carrying out their own projects to create a bureaucracy to weigh the relative merits of competing proposals. When choices had to be made, a top executive would sit down and say, "Okay, we've got this list. All right. I want to do that one and that one and that one. And the others, you guys wait. And that meeting would have been over and off we'd have gone." He recalled a meeting with one of the center executives one day, when a scientist walked in and announced that the center ought to be doing some work on aeronomy, the science dealing with the physics and chemistry of the upper atmosphere. "And Jack said, well, invent a project." The scientist came back a little later and said, "This is what I want to do." The executive approved the project.[41]

"Now you can't get permission to do a phase A study in less than six months," the science director complained. NASA inaugurated phase A studies to develop project ideas sufficiently to decide whether to undertake the projects or not. In part, the new procedures arose because there were so many good things to do and, as NASA matured, so little money with which to do them. "So then we get into competitions among ourselves as to which ones we should do. . . . You do studies and you study the living daylights out of things until such time as you can make the case that your project is better than somebody else's project because it has higher technical merit. And you've got to do a lot of that kind of justification. . . . You go to the National Academy of Sciences and you say, 'We'd like you to do a study and advise us which of these things are the ones that are best to do.' " The National Academy of Sciences has committees of scientists who can review NASA projects. "It is a lot harder to second guess an academy report," the science director observed, "than to criticize a NASA executive. Who is the executive?" he asked facetiously. "He's just the guy who makes decisions, you know."[42]

NASA employees typically blamed their loss of flexibility on tightening political control, imposed on NASA by people outside the organization. "There is a thing which I call bureaucratic decay," said one of NASA's first center directors. "It's a pretty virulent disease. And it comes about because of the politicization of the organization." Once a government agency starts to grow, the amount of political oversight invariably increases. "As soon as the NACA became NASA, it began to accrue." The NACA, he insisted, never became a political organization. "That's another thing about the NACA organization. It was always so small, it was hardly worth anybody's effort to do too much with it. Now of course it was protected at the top by [the advisory] committee. And the net effect was it just never got taken over that way."[43]

"NASA obviously is more bureaucratic now than it was in the very early days," said one of the agency's top space scientists. "It's just a function of growing older," he observed. "Look at the funding flexibility NASA had in the 1960s. The budget was equivalent to about $25 billion in today's dollars and there were billions of dollars of money that were appropriated years before they were spent. There was oversight and all that kind of stuff, but not the kind of thing that goes on today."[44]

Numerous factors worked to increase political oversight. The fire that killed three astronauts on Launch Pad 34 in early 1967 promoted closer congressional scrutiny. Congressional oversight into NASA program details increased considerably after the fire, and with it the NASA bureaucracy needed to respond to legislative concerns. At the same time, the number of legislative assistants working for members of Congress and available to make inquiries grew. In 1957, the year before NASA was formed, the 535 members of the House and Senate carried out their work with 4,500 staff assistants. By the mid-1980s, legislative staffs had grown to 15,000, a trend independent of NASA's growth and contraction. Additionally, congressional reforms instituted during the 1970s curtailed the power of committee chairs and democratized the work of committees. Such reforms restricted NASA's ability to handle legislative problems by working with a few powerful congressional leaders.

Staff growth was not confined to the Congress. The White House staff grew from four hundred employees in the 1950s to nearly eight hundred in the 1980s. In 1982, White House officials set up a Senior Interagency Group under the National Security Council to act as a clearinghouse for all space policy. It was replaced in 1989 with the National Space Council, another multimember body giving more outside officials the opportunity to review NASA programs.

The procurement regulations under which NASA was obliged to operate grew more complicated. According to one advisory committee, Congress enacted more than sixty new laws relating to procurement policy between 1965 and 1991. "In addition, 25 Executive Orders, 16 Office of Management and Budget Circulars and 24 Office of Federal Procurement Policy Letters have been issued, all of which affect the procurement process directly."[45]

The federal civil service agency restricted NASA's use of excepted and nonquota positions to attract top-rate professionals with higher pay. Congress designated those positions that could be filled at the highest pay, a decision that had previously been left to the discretion of the NASA Administrator. Managers at the NASA field centers scrapped to hold on to the special positions they already had. Bureaucrats at the Civil Service Commission insisted that NASA justify the contribution of its special positions to the objectives of President Lyndon Johnson's Great Society program.[46]

Other federal regulations swelled: accounting standards, affirmative action, requirements for peer review, occupational health and safety, environmental protection. These regulations spread not just to NASA but to all federal agencies carrying out work during that time. Individually, each change had a worthy objective that led to its enactment. Together, the changes produced an appalling growth of bureaucracy.

The loss of flexibility imposed by these developments was not something that NASA officials wished upon themselves. In fact, they deeply resented it, especially when it allowed what they viewed as interference in technical affairs. "The space station is a classic example," the space scientist continued. "You can't get around to building the space station because you are always busy doing another set of push-ups for some committee in town or some detailed oversight thing. You get some congressional staffer who is busy telling you which widget to put where, you almost get the impression they are going to tell you what size bolt holes to put on the thing. These people, some of them are technically competent, some are just not technically knowledgeable. It's a very constrained environment."

"I recently came to the conclusion that somebody ought to pass a law that made it a capital crime to do systems engineering in Washington." In the eyes of NASA professionals, systems engineering decisions are technical decisions best left to experts in the field. The government would save "a hell of a lot of money," the scientist said, if it pushed all of its technical decisions out to the field centers, where Washington insiders would have less opportunity to interfere in them. "There are too damn many experts in town helping to do the work, and they're not qualified to do it—but they are certainly in positions of authority to do it. Most of them aren't even elected officials or appointed officials, they're staffers." The real heroes of the space program, he concluded, were not the people who put Americans on the Moon. "The real heroes are the ones that have managed over the last twenty years to keep us in a position where we still have a civil space program after all the nonsense that goes on."[47]

Increasing Bureaucracy

A great deal of NASA's lost flexibility was imposed on the civilian space program by people outside of the agency. Much of it was imposed on the government as a whole, not just on NASA. In the beginning, NASA officials had the opportunity to resist bureaucratic intrusion into technical affairs. As NASA matured, they lost that capability. In response, the NASA bureaucracy grew. NASA's administrative apparatus became more complex. NASA needed more people to prepare and defend the budget, plan and defend new programs, meet with representatives from other agencies, respond to congressional inquiries, and deal with the White House staff. They needed people to coordinate complicated programs and monitor the work of contractors. They needed people to deal with civil service rules and procurement regulations. They needed people to review projects and allocate increasingly scarce funds.

In its early years, NASA acquired a reputation as a can-do agency, one in which employees could find a way to beat the government bureaucracy and get their projects done. That reputation changed as NASA matured. Outsiders interviewed for this study commonly expressed the view that NASA employees no longer viewed themselves as outlaws working to foil the bureaucracy but as agents of it. Lost flexibility and the inability of NASA employees to resist institutional growth conspired to produce a more bureaucratic space program as the civilian space agency matured.

Much bureaucratic growth in government is purposeful; that is, the people who run governments deliberately create it. As new problems arise, new regulations are written to deal with them. Citizens complain about the burden of governmental regulation, but the fact remains that most red tape can be traced to incidents for which the public has demanded a regulatory remedy.[48] In spite of the popular disdain for large bureaucracy, government leaders often find bureaucratic forms of organization preferable to the best alternatives. Machines (such as computers) are a substitute for bureaucracy; so are the use of private corporations to carry out public goals or the reliance upon organizations run by members of the professions. The public does not like to deal with machines, public officials are reluctant to contract out essential services like police or welfare, and politicians find professional organizations—in which doctors or engineers substitute their professional codes of behavior for written rules—hard to control.[49]

The bureaucratic form of organization maximizes predictability, objective treatment of clients, continuity, and material efficiency while giving politicians a fair measure of control over the work of the organization. Bureaucracies accomplish this by developing elaborate rules, maintaining a work force of career civil servants, vesting authority in a hierarchy of offices, and by other means. In government, it is a very popular organizational type. The tendency of politicians to select bureaucratic forms of organization has been well documented since the beginning of the twentieth century, when the German scholar Max Weber explained this emerging phenomenon.[50]

In addition to its purposeful functions, bureaucracy possesses a dark side. Much has been written about the propensity of bureaucrats to behave dysfunctionally, in ways contrary to the mission of their bureau. In one commonly observed dysfunction, bureaucrats switch means for ends, a phenomenon known in the scholarly literature as goal displacement.[51] The pressure to enforce rules becomes so strong in modern organizations that the rules take precedence over the goals for which they are created. In extreme circumstances, bureaucrats may fight to preserve outmoded rules, often by writing additional regulations to make old rules more binding. A presidential commission investigating the 1979 nuclear power plant accident at Three Mile Island criticized the U.S. Nuclear Regulatory Commission for displacing the objective of operational safety with a demand that the nuclear industry comply with ineffective rules.[52]

Bureaucracies are also thought to possess an uncontrollable tendency to grow. This tendency received the status accorded laws of nature through a half-humorous explanation by C. Northcote Parkinson. Parkinson's law states that "work expands so as to fill the time available for its completion." He offered as an example the number of employees in the British Colonial Office, which rose from 372 to 1,661 during an eighteen-year period when the British government divested itself of its largest colonies.[53] The evidence supporting this observation is not conclusive.[54] While government as a whole has grown, the size of individual agencies fluctuates over time.

Even if they do not grow larger, individual agencies do seem to grow more bureaucratic with time. Bureaucracy is a relative concept; agencies can possess it to varying degrees. Scholars have tried to explain the phenomenon of increasing bureaucracy in many different ways. Some suggest that the civil service attracts personalities with a predisposition for building bureaucratic empires. The evidence, however, seems to suggest the opposite—that the civil service attracts people who are flexible and idealistic and who dislike bureaucracy as much as the public at large.[55] Others point to forces outside of individual agencies—the growth in regulations and legislative oversight within the government as a whole. As the regulatory burden of the administrative state increases, so does the tendency of individual agencies to deploy more people into administrative posts to handle that burden.

Still others point to forces inside the agencies themselves, generally associated with the passage from youth to maturity. As government agencies mature, they formalize their methods of operation. They adopt management control procedures, often in response to crises that expose the absence of such measures. The situations with which officials deal grow more familiar. Rules are written to standardize the agency response to such situations. Some agencies alter their goals, shifting their attention from innovative work to maintenance functions that enhance the survival of the operation.[56]

Agency statistics and survey data support the notion of bureaucratic development. One of the earliest signs of impending bureaucracy is growth in an agency's central staff. As many old-time NASA employees observed, the NACA (NASA's predecessor) got along with a very small central staff. Between 1950 and 1956, the NACA headquarters staff fluctuated between 155 and 172 people—in an organization that housed 8,000 civil service employees. The central staff proved so unintrusive that NACA employees referred to it as the Washington staff or the Washington center, avoiding the more authoritative image implied by the term *headquarters.* From just 2 percent of the whole number of employees under the NACA, NASA's central staff grew to more than 2,000 employees, or 6 percent of the whole agency by 1963. As the agency cut back on the size of its operations during the 1970s, the proportion of people working in Washington crept toward 7 percent. It reached 8 percent in 1988, as NASA turned thirty years old. From the NACA to NASA, the central staff grew by a factor of ten in an organization that grew by a factor of three. Put another way, the Washington staff grew more than three times as fast as the agency as a whole. NASA employees no longer refer to the central corps as the Washington center. They now recognize it as NASA headquarters.

Some of that growth was imposed by the administrative requirements of more complicated programs, such as the central staff for Project Apollo. The NACA did not need a large central staff, because research projects could be delegated to single field centers. NASA executives split up their large space flight programs among many centers, thereby creating more need for central control. The need to respond to increasing political oversight and governmentwide administrative regulations also prompted the headquarters staff to grow.

The number and influence of professional administrators within NASA increased proportionately, another sign of increasing bureaucracy. NASA is staffed by scientists and engineers, technicians, trade and craft employees, clerical personnel, and professional administrators. Since the agency began, scientists and engineers have been the dominant professional group within NASA. Only at Washington headquar-

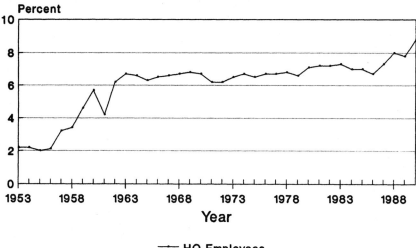

Figure 6   Percentage of NASA Employees at Headquarters

— HQ Employees

Sources: Alex Roland, *Model Research: The National Advisory Committee for Aeronautics, 1915–1958,* SP-4103 (Washington: NASA, 1985); NASA, *Pocket Statistics,* annual; and NASA Office of Human Resources and Education, "The Civil Service Work Force," annual (Washington).
Note: Percentage of permanent and temporary NASA employees at the end of the fiscal year. Figures for 1953 through 1958 are for the National Advisory Committee for Aeronautics.

ters, where policy planning and administration prevail, do professional administrators outnumber scientists and engineers.[57]

Within NASA, professional administrators are defined as people who perform "such activities as financial management, contracting, personnel, security, administration, law, [and] public affairs."[58] Their influence was so slight when the space program began that NASA did not even count such people in a separate category, lumping them together instead with clerks and secretaries. NASA officials began to count professional administrators as a separate occupational code in late 1960, when they made up just 5 percent of the agency's permanent employees. Between 1960 and 1967, the number of professional administrators expanded from 792 to 4,600, growing three times as fast as the agency as a whole. As NASA cut back the size of its work force between 1968 and 1985, the proportion of professional administrators grew again. This is consistent with the tendency of a maturing bureaucracy to rely

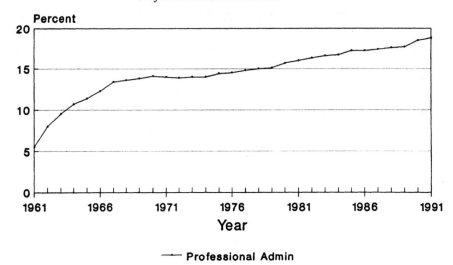

Figure 7   Percentage of NASA Employees Categorized as
Professional Administrators

Sources: Jane Van Nimmen and Leonard C. Bruno, *NASA Historical Data Book,* vol. 1, *NASA Resources, 1958–1968,* SP-4012 (Washington: NASA, 1988); and NASA Office of Human Resources and Education, "The Civil Service Work Force," annual (Washington).
Note: Percentage of permanent employees at the end of the fiscal year.

heavily upon administrative procedures to distribute resources and plan its survival during periods of contraction.

Professional administrators were more likely than NASA scientists and engineers to receive promotions during the period of retrenchment, another sign of their growing influence (see Figure 5). In 1969, the first year for which comparative statistics were kept, a person in the scientific and engineering professions was just as likely to receive a promotion as someone classified as a professional administrator. Members of both groups had about one chance in six of receiving a promotion that year. As the agency cut its employment rolls, opportunities for promotion fell dramatically. The promotion gap between professional administrators and scientists and engineers simultaneously widened, in favor of the administrators. In the depths of the retrenchment (1972), less than 3 percent of all NASA employees received promotions. A professional administrator, nonetheless, was *twice* as likely to receive a promotion that year as a NASA scientist or engineer.

The two-to-one promotion gap reappeared frequently during the

1970s and early 1980s. During this period, promotions for professional administrators, the bearers of bureaucratic norms, consistently exceeded promotions for scientists and engineers, ostensibly the dominant professional group within the agency and bearers of the technical culture. Professional administrators often have a greater desire for promotion, since a promotion upgrades their status by increasing the number of people reporting to them. From their positions on the central staff, professional administrators are often in a better position to control the distribution of promotions. They may use promotions as a means of increasing their pay when salary increases slow, a phenomenon known in government circles as "grade creep." Whatever the specific explanation, the growing influence of professional administrators lends credibility to the perception of increasing bureaucracy.

This perception of increasing bureaucracy was widespread among NASA's professional work force. As part of the survey undertaken for this study in 1988, NASA professional employees (engineers, scientists, and professional administrators) were asked to assess the extent of increased bureaucracy. Eighty-four percent of the respondents agreed that "NASA places a great deal of emphasis on paperwork and procedures." Only 11 percent thought that "it is relatively easy to cut through the bureaucracy and get things done within NASA today." Only 9 percent thought that "it is fairly easy to change official procedures within NASA once they are approved."

To gauge the extent to which the bureaucracy grew as the agency matured, respondents were asked the degree to which practices had changed since they joined NASA. Employees who worked in NASA before 1970—49 percent of the total group—were more conscious of change than those who joined in 1970 and beyond. Employees from NASA's early years were more conscious of increased paperwork and the inability "to cut through bureaucratic barriers and get things done" (see the appendix: the NASA bureaucracy).

What people outside NASA often viewed as excessive bureaucracy, some inside NASA viewed as necessary to their job. Outsiders complained about the bureaucratic burden NASA imposed on its contractors. Indeed, NASA invested more effort in contract oversight than other government institutions like the U.S. Air Force.[59] The paperwork on a large NASA contract could be as demanding as the task of actually building the hardware. Contractors viewed the paperwork as an excessive burden, but NASA professionals viewed it as contractor penetration—an essential element in their overall culture.

On the whole, however, bureaucracy proved no more popular

within NASA than it was among the public at large. It was typically seen as a challenge to the flexibility and innovation needed to maintain NASA's technical culture. "Instant bureaucracy," one NASA executive called it. NASA was "getting itself so bound up in paper that [it was becoming] a substitute for thinking."[60]

Growing More Conservative

In addition to becoming more bureaucratic, the behavior of agency officials is thought to become observably more conservative as the agency matures. This can manifest itself in a number of ways. The people who lead the organization may become less willing to take risks or endure failure. Employees may become distrustful of change. Top officials may be less receptive to new ideas, or they may grow reluctant to take actions that would disturb the balance of power that supports agency affairs. Whatever the path, government institutions founded as agents of reform often can be found decades later in the embrace of organizational conservatives.

Numerous examples of organizational conservatism in government agencies can be found. During the Great Depression, politicians praised the accomplishments of the Tennessee Valley Authority with much the same rhetoric later devolved upon the NASA space program. The TVA provided a dramatic symbol of U.S. technology and national will. Fifty years of increasing bureaucracy and sunk costs transformed the TVA into what one commentator called "the very type of organization it set out to challenge."[61] Similar accusations have been leveled at the Federal Trade Commission, the Federal Maritime Commission, and the Federal Food and Drug Administration.[62]

Social scientists offer various theories to both confirm and explain this tendency. Anthony Downs, in his classic work on the life cycle of bureaus, offers an explanation based on the types of personalities that inhabit institutions during the various stages of growth. A young, expanding agency attracts people who are more inclined to take risks and embrace new ideas, Downs suggests. Downs characterizes them as climbers, zealots, and advocates for the agency's cause. They dominate the management of the agency until its period of growth ends, at which point diminished resources and fewer promotion opportunities prompt the climbers to leave. Their place is taken by people that Downs characterizes as conservers, people who "seek merely to retain the amount of power, income, and prestige they already have."[63] Risk takers who stay tend to lower their expectations and become conservers themselves.

"All organizations tend to become more conservative as they get older," Downs predicts.[64] The bureaucratic growth that comes with age also favors the power of conservers.

Other social scientists blame conservative behavior on the tendency of government agencies to become excessively dependent on their clientele. All government agencies in the U.S. government need to develop a clientele. Without one, the agency's political base withers, along with the budget its leaders can expect. Some agencies officially involve their clientele in the administration of agency programs. The U.S. Department of Agriculture, for example, relies upon a network of local advisory committees to assist departmental executives in the distribution of farm loans and crop subsidies. Some agencies seek to co-opt their clientele, allowing special interests the opportunity to influence agency policy in exchange for political support.

Clientele so embraced may gain the upper hand in the relationship with their benefactor. The agency may come to rely upon information provided by the clientele to set policy. Agencies dispensing grants or contracts may find themselves locked into programs upon which their clients depend. Some observers suggest that agencies thus affected become captured by their clientele. In practice, the relationship is often more subtle than that. The clientele does not need to control agency policy and personnel in order to promote conservative behavior. It merely needs to make the agency more dependent upon the information and support that the clientele can provide.[65]

Economists offer even another explanation for increasing conservatism. Public executives, some assert, grow risk-averse because they are not required to compete with other institutions in the same way that business managers in a competitive market do. Business managers must take risks and adopt new ideas in order to survive.[66] Public bureaus, on the other hand, are typically insulated from competition. They receive their revenues from a single customer (the legislature) and commonly enjoy monopoly status in the delivery of their services. In his book *Bureaucracy and Representative Government*, William Niskanen attempts to demonstrate how public bureaus so situated will grow faster and be less willing to adopt new technologies than firms fighting for survival in the marketplace.[67]

The NASA experience shows how an agency can grow more conservative even as it embraces a culture of innovation. NASA possesses an organizational culture that prizes change. It is, after all, a research and development organization. Through space exploration and flight research, NASA employees seek to create technological change. They have witnessed the virtues of innovation and the rewards of progress. They have publicized the technology spin-offs that space exploration and flight research create.[68] They emphasize the need for risk taking. They want to explore new frontiers.

While all these things are true, it is also true that NASA employees worry about their organization growing more conservative. Much of this is based on the perception that the NASA leadership during the second and third decades of space flight became less willing to take risks, often as a result of forces beyond their control.

A conservative organization, by definition, is one in which employees do not welcome change. Under the terms of NASA's original culture, however, agency employees were encouraged to seek out new frontiers. This attitude persisted as NASA matured. When asked if "NASA should concentrate on implementing programs already approved rather than pushing for new programs," only 20 percent of NASA professional employees responding to the 1988 culture survey said yes. The majority wanted to push ahead to new frontiers, as the original culture prescribed. As for NASA's role in pursuing those frontiers, 53 percent agreed that they were "very optimistic about NASA's future." NASA's future, as described in the agency's long-range plans, contained daring expeditions and wondrous machines. NASA professional employees hoped that those projects would be approved and become part of the national agenda.

When asked about the possibilities for actually managing that change, attitudes shifted dramatically. NASA professional employees were asked whether they agreed with the statement that "most of the changes that I have seen take place within NASA have made the agency work better." Fifty-six percent disagreed. Only 18 percent thought that the changes they had witnessed had worked out. People who had served in NASA during the first generation were more pessimistic about the implementation of change than recent recruits.

When asked directly whether "people within NASA welcome change," respondents to the survey were equally divided. Thirty-eight percent said they did; 37 percent said no. Interestingly, people who arrived at NASA after the landing on the Moon were more pessimistic on this point than the members of the Apollo generation (see the appendix: attitudes toward change).

Such ambivalence toward change is consistent with the competing pressures created by a technical culture amid growing bureaucracy. The technical culture encourages employees to be optimistic about the future and the gains to be realized from scientific and technical change;

experience with a sluggish bureaucracy teaches employees to be wary about the possibility for making the changes work from within. In their thirty-year history, NASA employees did not come to the point of rejecting change, but they did grow more pessimistic about how much good would come from pursuing it through their own organization.

People lose faith in the ability of their organization to manage change as the organization grows rigid. A conservative organization is one in which conservers, whose outlook encourages them to resist change, come to dominate overall management. Such people make change difficult though excessive caution, by raising procedural roadblocks, and by frustrating the advocates of new ideas. They discourage risk taking. They insist upon adherence to official procedures, a development that constricts the ability of the institution to attain the high levels of flexibility and rapid communication necessary to make complex programs work.

NASA professional employees agree that such things have taken place within their agency. Among those responding in the 1988 survey, 62 percent agreed that "at the management level, NASA is dominated by people who are cautious and inclined to avoid risks." Only 17 percent disagreed. Survey participants were also asked whether "the number of people who are cautious and inclined to avoid risks has increased since I joined the agency." A majority agreed. The perception of creeping conservatism was especially strong among the people who joined the civilian space program before 1970. Sixty-six percent of the first generation believed that NASA management had grown more conservative, compared to 42 percent from the second generation (see the appendix: managerial risk taking and communication). A complementary study of NASA's organizational culture by Warner Burke confirmed significant concern about the willingness of senior managers "to do the right thing." Although individual employees expressed optimism about the future and valued high-quality work, they worried that agency management would base decisions "on political concerns as opposed to research and technical criteria."[69]

One of the more important measures of organizational flexibility is the degree to which employees are willing to communicate with each other. In rigid bureaucracies, employees avoid face-to-face encounters and use official procedures as a means of avoiding open communication. For them, bureaucracy serves a positive function. It provides them with an impersonal means for making what would otherwise be personal decisions. It allows them to fall back on a rule instead of having to deal with people.[70] NASA professional employees agreed that their colleagues did not "communicate with each other as much as they should," by a margin of 65 to 18 percent. Communication breakdowns were identified as a contributing cause of the space shuttle *Challenger* accident, a concern repeated four years later in the report of the Augustine committee investigating the future of the U.S. space program.[71]

A conservative organization is also one in which people learn to take fewer risks. Risk taking has many meanings within NASA. Employees take risks when they advocate new ideas. They take risks when they phase out old programs and start up new endeavors. They take risks when they put new technologies in their spacecraft and satellites. They take risks when they conduct flight tests, especially with astronauts and pilots on board. They take risks when they explore new frontiers.

Thirty years after its founding, risk taking remained as a very important part of NASA's culture. NASA professional employees continued to believe that "risk and failure are a normal part of the business of developing new technologies." Practically no one, including the professional administrators who answered the survey, disagreed with this basic belief. Putting belief into practice, however, apparently proved harder to do, as the responses to three follow-up questions revealed. More people than not believed that "NASA employees are allowed to fail and learn from their mistakes," but a substantial number (31 percent) believed this was not true. More people than not agreed that "NASA rewards people who are technically creative," but a substantial number (31 percent) disagreed. More people agreed that "NASA has stayed on the cutting edge of new technologies," but 35 percent disagreed. In each case, a sizable minority expressed discontent about the allowance for risk taking, a concern equally distributed in most cases among early and recent recruits to the space program (see the appendix: attitudes toward risk and failure).

The NASA experience suggests that it is possible for members of an agency to dislike bureaucracy as they grow more bureaucratic, to praise the virtues of change while watching their agency grow more conservative, to believe in taking risks yet feel less able to take them. What NASA employees experienced as the agency matured were competing cultures, or at least the pressures imposed by bureaucratic weight on the technically rooted beliefs of the people who made up the organization. In many cases, the people who complained about NASA's loss of flexibility were the officials who actually ran the space program. They blamed forces over which they felt they had little control. "There was a very sharp increase in the bureaucracy immediately following the 204 fire," said one NASA scientist, expressing a common point of view.[72] Up until that time, NASA officials gave their spacecraft managers a relatively free hand in carrying out their responsibilities.[73] To avoid future disasters like the 204 fire, top officials increased oversight and tightened up bureaucratic control. "After that," said the scientist, "there was a gradual attrition of the competent risk-takers and a growth of conservatives."[74]

Failure apparently taught NASA officials to become more conservative. Outside experts familiar with the NASA space program frequently offered such observations when interviewed. When the first six Ranger probes to the Moon failed and Congress ordered a legislative investigation, NASA officials became less willing to take risks on the following flights. NASA officials began planning for success. They built in protective layers of administration and lots of contractor oversight. They became more conservative, more concerned with not making mistakes, and less tolerant of intuitive decisions. The Apollo fire and the *Challenger* accident magnified this tendency.

NASA managers who agreed with this interpretation nonetheless continued to believe in the old cultural value of risk taking. Their technical instincts taught them to value change; their organizational experience taught them to be wary of it. They did not want to become conservative but believed that the political system forced them to. "There is too much scratching around for problems by the press and the Congress," noted one headquarters official. "We are extremely open and susceptible to second-guessing and oversight. It's gotten so bad that if an issue comes up the first thing you have to ask is what will the press do with this or the congressional staff. They can kill it before you get a chance to talk about it."

The inquiries and NASA bashing that followed the revelation of a problem bothered the official, because in his view it was prompting a shutdown in communication. "You can't talk about anything bad or it will hit the press. . . . You can hardly whisper it in the dead of night. That is really bothersome to me, because it's going to close down the agency. It's going to make you terribly conservative, and I don't think it's us. That's the problem. I don't think it is NASA that wants to do that." NASA, in his view, was being forced to be conservative. "It will change the culture so much that we may well not be the same agency that we were." Creative, innovative people, he observed, did not want to work for an agency where they had to measure the political implications of every word they uttered. "They are going to say, 'The hell with this, we'll go somewhere else,' and pretty soon you've got pretty mediocre people. As soon as you've got that situation you're not going to be out there at the cutting edge and you can easily become an Amtrack of space," operating an outmoded and inefficient transportation system.[75]

Another NASA insider bemoaned the effect of increased political oversight on the civilian space program. "In the sixties the ability of an individual to influence programs and the agency itself without being challenged by the OMB and half a dozen other agencies around town and the Congress and one's supervisor and five other associate administrators and seven staff people was considerably different than it is today." Even though the executive had moved to the top of the NASA civil service hierarchy, he felt that his influence over the space program had been diminished by increasing oversight. "I don't know when in my career I felt that I had the greatest influence on what the agency was doing," he said, "but it is not today."[76]

Fighting for Survival

Survival struggles have a profound effect upon public administration. Common wisdom holds that government bureaucracies rarely die, even after the purpose motivating their creation has disappeared.[77] While survival is the general rule, a smooth route to survival is not. Few public institutions last long without enduring episodes during which their existence is at stake. To survive such trials, agency executives frequently modify the services they provide or their methods of doing business. Institutions so affected may emerge from these incidents with significantly altered capabilities.

Governmental cutbacks during the 1980s rekindled scholarly interest in the effect of survival episodes on government operations.[78] Do threats to survival encourage organizational innovation or do they give rise to rigidity?[79] One might expect that a fight for survival would prompt the leaders of large institutions to embrace new ideas. Institutional survival in the public sector, after all, resembles competition in the business world. Public agencies struggle for a share of a public purse of limited size. One might also anticipate that the administrative apparatus of a shrinking agency would contract, making the agency leaner and less bureaucratic. This is not typically the case, however, for agencies on the downward slope of their growth and contraction curve. While public executives do exhibit a great deal of creativity in responding to decline, technical innovation and rapid restructuring are not typically part of their repertoire.[80]

Scholars explain why survival efforts in public administration often give rise to rigidity and hypercomplexity. Government executives faced with the problem of diminishing resources are frequently obliged to devote more efforts to administrative functions, like planning. Precisely because their resources are diminished, plans and procedures for distributing what remains must be more carefully constructed so as to guard against the claims of the disappointed. Executives who preside over a shrinking organization with many parts—such as an agency with installations in the field—may distribute the programs that remain by slicing them up in odd ways. While the parcelling of small pieces among many installations enhances the survival of each, it also complicates the problem of overall coordination. With coordination problems on the rise, agency chiefs have to enlarge their administrative staffs. This is one important reason why a shrinking organization may find its administrative apparatus in a state of growth rather than contraction.

Government agencies are often obliged to shift the purposes they satisfy in order to survive. Routine operations frequently increase in importance in such organizations, simply because the funding and management of routine operations are more predictable. All organizations abhor uncertainty, and public executives engaged in a fight for survival can ill afford to increase the already heavy amount of uncertainty with which they must contend.[81] Routine operations provide a safe haven for such organizations. Agency leaders may seek to acquire predictable sources of revenues through the earmarking of taxes, for example, or by incurring obligations without appropriations.[82] They may seek to reduce the impact of criticism by acquiring monopoly status over the delivery of a needed service.[83] The activities that emerge can prove a poor match to the original purpose of the agency, but they can enhance its survival.

There is no question but that employees at NASA installations felt threatened during the downturn in space activities during the 1970s. The Manned Spacecraft Center (now the Johnson Space Center) was built on land transferred from an American oil company to the U.S. government through Rice University. An oft-repeated rumor held that the deed of transfer required the government to return the land to the university should the government decide to terminate human space flights after Americans landed on the Moon. While the rumor was not true, it fed fears that the government planned to close the Manned Spacecraft Center once the lunar expeditions drew to a close.[84] Twice during the 1970s the government conducted studies that could have led to the closing of the Marshall Space Flight Center in Huntsville, Alabama.[85] "We came very near to it," said one center director, "nearer than most people know."[86]

With more and more of NASA's research being performed under contract, overseers of the space program wanted to know why NASA needed to maintain three in-house research laboratories. The Lewis Research Center "barely escaped closing," according to one analyst, and the Ames Research Center came under scrutiny as well.[87]

NASA officials exhibited a good deal of flexibility in promoting the survival of these centers. Some centers diversified. At the Marshall Center, engineers who had built their reputation in the area of rocketry moved into the space station and space platform business. Officials at the Marshall Center took on the responsibility for developing *Skylab*, America's first orbital workshop. They were responsible for practically every part of *Skylab* except the astronauts and their spacesuits.[88] They also took charge of the development of the Hubble Space Telescope.[89] They developed plans for automated space platforms, pallets that could carry instruments in low earth orbit and be transformed into space stations with the addition of habitation modules.[90]

Managers at the Lewis Research Center, encouraged by high-level officials who wanted NASA to address earth-based problems, searched for clients outside NASA to whom they could offer their work. They moved their research program back into aeronautics, where they had both military and commercial clientele. They developed instruments to monitor air quality in their hometown of Cleveland, Ohio. They won funding from the Environmental Protection Agency to develop a gas turbine automobile engine. Electricity-generating windmills and coal-burning power plants also showed up on the Lewis research agenda, as the center diversified.[91]

Even NASA headquarters exhibited flexibility, reshaping its programs to accommodate the desires of White House and congressional officials who allocated program funds. Between 1969 and 1972, they modified their design of the Space Transportation System and offered to reduce its size in order to get it approved. Between 1985 and 1991, they redesigned space station *Freedom* no less than five times in their efforts to find a configuration that the politicians would fund.

Survival efforts proved to be a two-edged sword. While such efforts fostered flexibility in some ways, particularly in the search for new functions to perform, they restricted flexibility in others. Comments about NASA's loss of flexibility were most commonly made by outside experts commenting on NASA operations.

Diversification apparently restricted administrative flexibility. By permitting field centers to diversify into each other's business, NASA officials increased requirements for bureaucratic overhead and program coordination. Programs run by two or more centers are harder to coordinate than those run by one. Development of the Hubble Space Telescope illustrates this fact. Officials at both the Marshall Center and the Goddard Space Flight Center competed for the right to develop the telescope. The Goddard Center drew its expertise in space observatories from its early satellite program, especially the Orbiting Astronomical Observatories. The Marshall Center based its expertise on the *Skylab* facility, which contained a solar telescope, and on the fact that manned flight activities would be needed to deliver and service the telescope. NASA executives finally worked out a memorandum of agreement between the two centers whereby the Marshall Center would have overall authority for building the telescope (the work actually to be done by twenty-six major contractors and subcontractors) and the Goddard Center would be responsible for instruments and flight operations.[92] By giving each center an equal share of responsibility, NASA officials created management problems in the large telescope program that were never fully resolved.

Work on the space station *Freedom* was divided up among four NASA field centers. The Johnson Space Center was allowed to develop the structural framework that held the space station together, the Marshall Center received responsibility for the modules making up the station core, the Lewis Center won the right to develop the station's power system, and the Goddard Center took charge of the free-flying platforms.[93] Additionally, the Johnson Center was given overall responsibility for coordinating the four centers, including itself. Under this complicated scheme, it was nearly impossible to tell where one center's responsibility ended and another's responsibility began. "There were some bitter battles about where that dividing line was," said one of the top program officials.[94] The management system collapsed two years after its inception.[95]

By the necessity of having to protect field installations whose programs were threatened, NASA officials adopted policies that made administration more difficult. Commenting on NASA's management problems, a special advisory committee of outside experts chaired by Norman R. Augustine suggested that "projects appear to have been tailored to help perpetuate the work force, rather than the work force having been tailored to meet the needs of the project." While diversification may have helped NASA centers to survive, it also maximized the number of organizations whose cooperation was required in order to make a project work. This, the Augustine Committee argued, was "exactly the opposite of generally accepted management philosophy that argues for minimizing interfaces, the 'nooks and crannies' where problems seem to breed."[96]

Survival struggles restricted flexibility in a second way, in the receptivity of the agency to new ideas. This observation was also most often voiced by people outside of NASA. NASA employees, by a fairly narrow margin, expressed the belief that the agency "has stayed on the cutting edge of new technologies" (see the appendix: attitudes toward risk and failure). Outsiders familiar with agency operations were not so convinced, especially those who worked in institutions that competed in the marketplace.

"Among the concerns that have been most often heard," reported the Augustine Committee, "has been the suggestion that the civil space program has gradually become afflicted with some of the same ailments that are found in many other large, mature institutions, particularly those institutions which have no direct and immediate competition to stimulate change."[97] Business firms or privately financed research institutions that compete for customers and funds cannot afford to let a competitor beat them to a new technology. Their continuance is at stake. Without new ideas, in a competitive market, they cannot survive.

The situation in government is substantially different. Government agencies compete for funds, but they do so against agencies that provide entirely different services. The NASA budget is reviewed by House and Senate appropriations subcommittees that also review funds for housing, veterans, and environmental protection. In such a political environment, new ways of doing business do not give a government agency a significant advantage in the competition for funds. In fact, new ideas can often weaken the agency's position. Old ideas are funded without much investigation as part of the agency's on-going base, while new ideas receive a large amount of scrutiny.[98] New ideas have to wait in line, a risky proposition for agencies fighting for survival.

Agencies struggling to survive sink funds into established programs that still command political support. Since the number of programs that enjoy political support typically exceeds the capacity of the government to fund them, the agency acquires backlogs. The capacity of agencies so affected to move into entirely new endeavors is severely restricted.

In 1991, NASA officials pressed hard to win approval to build another space shuttle orbiter. Critics called it an old idea. By the year 2000, the space shuttle would be a thirty-year-old technology. Many advances in flight control and computer technology would have been made during that time. In 1989 President George Bush directed NASA officials to start making plans for a lunar base and an expedition to Mars. The NASA scenarios for the Moon/Mars expeditions seemed so conservative to officials on the National Space Council that they asked a panel of outside experts to search for new ideas.[99] Concern over NASA's inability to move away from its established agenda and undertake new initiatives contributed to the White House request in 1992 for the resignation of the NASA Administrator.[100]

Do threats to survival encourage agency leaders to innovate and adopt cost-saving measures? Or do they discourage new ideas? The scholarly literature on organizational decline favors the latter, although no consensus prevails.[101] The NASA experience suggests that government agencies may be quick to diversify and seek out new functions, but that in so doing they may complicate their administrative burden and create a backlog of priorities that affects their ability to adopt new ideas.

Weakening Organization

As NASA changed, the organizational forces that once served to unify the agency during the Apollo program became weaker. During the 1960s, the people who came to Washington to manage Project Apollo centralized the job of program management. After they left, NASA executives decentralized the job of coordinating the next two big space flight programs—development of the space transportation system and construction of space station *Freedom*. They did this by lodging overall program management responsibility for these two initiatives at a single center in the field, what was known within NASA as the "lead center" approach.

Traditionally, leaders of the NASA field installations have resisted the efforts of people at headquarters to manage center affairs. Field center employees view their installations as little NASAs and work to acquire control over the projects they manage. Some even view their field centers as NASA. This is a normal part of the technical culture. By encouraging local responsibility, decentralization helps to promote technical capability. For self-contained projects that involve a small number of contractors, it is not an unacceptable way of doing business.

For the large Project Apollo, however, that tendency had to be counter-balanced by a strong program office in Washington, D.C. The leaders of Project Apollo coordinated and scheduled the work of many centers and contractors contributing to the expedition to the Moon.

To develop the Space Transportation System and the space station *Freedom*, NASA officials had to coordinate the work of many centers and contractors too. Agency officials, however, abandoned the central-ized program office as the means for doing this. The primacy of the mission no longer commanded a balance between central management and technical capability, especially as the space program struggled to survive. One of the leaders of Project Apollo left NASA as the lunar expeditions began in 1969, taking assignments in other government agencies and in the aerospace industry. In 1986 he returned in an advi-sory capacity. He recalled his impressions after seventeen years away:

"I was frankly surprised at the extent to which practices and disci-plines that had been, I thought, pretty soundly established within NASA had fallen into disuse." Some practices remained, but many had disap-peared. The Rogers Commission investigating the space shuttle *Chal-lenger* accident, he noted, "was quite critical of the way in which the shuttle program was being managed." The shuttle program manager at the Johnson Space Center did not have enough authority to control the work of other centers contributing to the program. Officials at the Marshall Space Flight Center tended to hold back problems and did not report them to the shuttle program manager or to flight readiness review teams. The management tools available to the program manager for Project Apollo some twenty years earlier—planning and control meth-ods and technical controls—were not being used by shuttle manag-ers.[102] These tools, the Apollo executive continued, were "just the fun-damental way we ran the Apollo program. They were, I thought, fairly well-established within NASA back by the end of '69. Many of them had been discarded."

"I could hardly believe what I found in the space station program," the executive said. Once again, NASA officials had put the program manager at the Johnson Space Center, reporting to the center director, and then told that person to coordinate all of the centers contributing to the program as well as the foreign nations participating in it. "It was just crystal clear that if the space station program was going to have any chance at all of succeeding, somebody was going to have to be put in charge and supported adequately by a program office and a systems engineering organization that could plan and direct that program." Field officials, preoccupied with the buildup of their centers, were extremely reluctant to place themselves under such a center of authority, as had been done for Project Apollo. They much preferred the low level of supervision created by the lead center approach.[103]

"I could hardly believe that the pattern of appropriately authoritarian responsibility and control of programs in general and especially the large programs had largely disappeared," the Apollo executive observed. "The space station tended to look much more like a save NASA program than a national space station program. It tended to look much more like a way to divide a multibillion-dollar federal multiyear budget in a way to preserve the survival of eight or so NASA centers than a logically established program to create a national space station." The lead center approach did not work any better for the space shuttle program. "The JSC was the lead center, but they weren't in charge of that program," the Apollo executive observed. "The people in the other centers wouldn't even give Houston the courtesy of responding to their communications, much less recognizing them as being in charge."[104]

Another leader of the Apollo program questioned the wisdom of the lead center approach when it was first proposed. "I said, I literally think we have changed the most successful management program this country has ever seen. You know, the new program technique might work, but if it doesn't, I do not know how you are going to explain it to a country that thinks Apollo and NASA synonymously in one word represent the epitome of management efficiency and risk taking."[105]

"The culture changed," another Apollo executive explained. It changed back to what it had been prior to NASA, when the NACA research laboratories enjoyed significant autonomy. "In the research centers," he said, "we want to have them independent." What was good for the research centers, however, was not applicable to large multicenter space exploration programs. That was why his generation of executives had acted "to very strongly centralize" the management of Project Apollo. "If you have a multicenter program large enough to involve a number of centers, then you need to have some control or else you just can't get the kind of visibility and the kind of communication setup that will permit you to carry out that program efficiently and effectively."[106]

The advocates of decentralization viewed local control as a means of reviving NASA's technical culture by shifting the integration work on complex programs back to the field. In fact, decentralization became a means by which people in the field elevated bureaucratic barriers to cooperation and communication.[107]

NASA's maturation in its second and third decades of work conformed to many but not all of the predictions contained in various life cycle theories of government organization. The agency lost flexibility and grew more bureaucratic, although much of that change was generated by the bureaucratization of the government as a whole, a trend over which NASA executives had little control. Agency management grew more conservative and risk-averse, but agency employees retained their commitment to the traditional culture of change. Agency executives creatively sought out new programs in their efforts to make NASA survive, but in consequence created an overly complicated administrative system and an agency with a reduced capacity to accept new ideas. The centralizing forces imposed by the requirements of the Apollo program were replaced by decentralized methods that permitted cooperation barriers to arise.

NASA as it matured came to resemble the more conventional government bureaucracies from which it had once distinguished itself. The agency, in the words of one employee, became "mired in mediocrity." NASA, he added, "is truly the Post Office and I.R.S. [Internal Revenue Service] gone to space."[108] As bureaucratic practices spread through the agency, the original technical culture that had once served as a source of resistance to bureaucracy began to lose its vitality.

5 ▲ Losing the Technical Culture

Heat cannot be transferred by any continuous,
self-sustaining process from a colder to a hotter body.
—The second law of thermodynamics

During the first decade of space flight, a strong technical culture guided the work of NASA employees. The norms typical of that period required NASA to maintain a corps of professional employees deeply involved in the details of space flight and aeronautics. The technical culture counterbalanced many organizational forces that rose up to challenge it. It overpowered the usual bureaucratic tendencies present in government operations. It provided a counterweight to the centralizing and organizational necessities of the Apollo mission.

During the second and third decades of space flight, the force of NASA's technical culture declined. The tradition of in-house technical capability gave way to the emphasis on contracting out that had been present since the beginning of the space program. Space flight became more continuous and repetitive, undercutting the original philosophy that treated space travel as a unique and inherently risky business. NASA officials did not do as much flight testing as they had done during the first decade, thereby altering the culture of verification. Tolerance for risk and failure diminished, even as the agency continued to operate risky systems. As these changes occurred, the job of the NASA civil servant became less technical and more administrative. New practices did not require the same level of technical involvement as the old. NASA employees found themselves increasingly distanced from the actual work of space flight, while organizational resistance to bureaucracy diminished. The original culture lost its power to elicit behavior compatible with the dominant cultural norms.

Contracting Out

NASA's technical culture was significantly transformed by the triumph of contracting out. The diminished importance of in-house work occasioned by the new philosophy compromised the agency's technical capability, reduced the opportunity for hands-on work, and turned NASA scientists and engineers into contract administrators. No single factor affected NASA's technical culture more than the increased use of contractors.

Extensive contracting was not part of the technical culture that NASA inherited from the NACA and the ABMA. The inherited culture encouraged its advocates to perform their most important functions in-house and maintain technical shops to support that capability. NACA executives insisted that their own engineers and technicians conduct the research that was the agency's primary product. ABMA rocketeers maintained tight control over their own test activities and retained the capability to build their own rockets on the grounds of the Redstone Arsenal. Where the NACA and the ABMA employed contractors, as in the construction of facilities, it was always in a supporting and not a controlling role.

NASA executives came under pressure to increase their use of contractors as soon as the agency was formed. To carry out Project Mercury, NASA executives enlisted the support of private industry. They also contracted services and hardware from other government agencies. The newly created space agency procured Mercury space capsules from the McDonnell Aircraft Corporation and Redstone launch vehicles from the ABMA (still part of the U.S. Army), which ordered them from the Chrysler Corporation. For the first orbital flights, NASA executives obtained their Atlas launch vehicles from the U.S. Air Force, which received them from the Convair Corporation. NASA project managers procured launch services from the U.S. Air Force, which oversaw a collection of organizations anxious to use the Atlantic Missile Range at Cape Canaveral. NASA officials set up a worldwide tracking system and coordinated recovery operations with the assistance of the U.S. Navy. Subcontractors assisted contractors. The industry-agency relationship occasioned by Project Mercury severely strained the NACA tradition of the agency engineer in charge using in-house facilities to complete whole projects.[1]

With the advent of Project Apollo, NASA's reliance upon contractors increased even more. Contractors took over much of the testing function. The decision to go to all-up testing of the Saturn launch vehicle and Apollo spacecraft meant that NASA employees could no longer perform step-by-step qualification flights for each element of the mission. Instead, contractors were told to give the spacecraft and rocket components a thorough testing at the manufacturing plant and deliver them to the launch site ready to fly. Testing responsibilities shifted away from NASA toward its contractors.

NASA executives made a deliberate decision to build up industrial and university capability to do work in space, using government contracts.[2] It broadened American technical expertise. It allowed NASA to organize the lunar expedition much more rapidly than if agency executives had tried to do this work in-house, waiting for Congress to approve the necessary facilities and personnel. Contracting out also broadened NASA's political base, creating a state-by-state constituency with a direct stake in an aggressive space program. By the mid-1960s, with the Apollo space program in high gear, NASA was sending more than 90 percent of its $5 billion annual budget to outside contractors and other government agencies. In 1960, NASA employed 36,000 contractors—three contract employees for every civil servant on the NASA payroll. By 1965, at the height of the Apollo mobilization, more than 375,000 contract employees worked on various NASA programs—ten contract employees for every in-house employee.[3]

Most NASA employees understood that the agency needed contractors to help carry out the human space flight programs of the 1960s. NASA needed contractors to fabricate spacecraft and rockets in order to meet the end of the decade deadline for lunar landing. That still left plenty of room for in-house traditions, however. NASA employees wanted to continue the practice of building satellites and smaller spacecraft in-house. They wanted to conduct much of their research activities in-house, as they had done in NACA. They wanted to maintain their workshops and build up the laboratories that provided project managers with technical support. They insisted on providing functions like spacecraft control and mission planning in-house. They wanted government employees to perform their own service activities, like spacecraft tracking and computer programming. NASA officials of the first generation treated those occasions where they had to contract for services as, in the words of one analyst, "occasional departures from the general norm that government provide such services in-house."[4] In programs like Project Apollo where contractors did most of the work, NASA managers developed a tradition of contractor penetration. Agency chiefs took great pride in their technical capability and did not hesitate to tell industry exactly what they wanted.

Table 4   Distribution of Permanent NASA Employees:
Five Bellwether Years

	Number Employed					
Year	Scientists and Engineers	Technical Support	Trade and Craft	Clerical	Professional Administrators	Total
1958	2,648	712	3,474	1,033	a	7,867
1962	7,982	2,644	6,182	3,489	1,755	22,052
1966	13,556	4,429	5,566	5,871	4,116	33,538
1983	11,094	2,631	1,329	2,887	3,562	21,505
1989	12,775	2,401	906	2,852	4,085	23,019

Sources: Jane Van Nimmen and Leonard C. Bruno, NASA Historical Data Book, vol. 1, NASA Resources 1958–1968, SP-4012 (Washington: NASA, 1988); NASA Office of Human Resources and Education, "The Civil Service Work Force," annual (Washington).
aCombined with clerical category prior to December 31, 1960.

Work force statistics show how NASA officials tried to retain in-house capability even as they contracted out 90 percent of their budget. The total number of technical support and trade and craft employees working inside NASA grew steadily during the early years. These were the people who ran the shops and built the rockets and provided NASA with its technical base for in-house capability. Between 1958 and 1966, the number of technical support and trade and craft employees more than doubled. It did not grow as fast as the agency as a whole, but it grew none the less.

This situation changed once the expansion of the space program ended and NASA entered its period of contraction. In reducing its size, NASA did not return down the path along which it had come. It backed down a different path. Trade and craft employees absorbed a disproportionate share of the cutbacks. NASA executives retained on the government payroll more scientists, engineers, and professional administrators than if NASA had simply run the pattern of growth in reverse. More scientists and engineers remained, but they had far fewer trade and craft employees to help them perform in-house work. By 1983, NASA had shrunk back to the size that it had been when it was just four years old. The distribution of employees in 1983, however, was quite different than it had been when the agency was young.

By the 1980s, NASA's contraction phase had ended. NASA budgets even expanded some. Rather than direct those increases into in-house capability, however, NASA executives watched as the funds went out to contractors for a wide variety of services. Over 90 percent of the increase in NASA's budget between 1978 and 1989 went to contractors.

Table 5   Percentage Distribution of Permanent NASA Employees

Year	Scientists and Engineers	Technical Support	Trade and Craft	Clerical	Professional Administrators
1958	33.7	9.1	44.2	13.1	a
1961	34.3	11.8	34.3	14.1	5.5
1964	39.2	11.7	21.4	17.0	10.7
1967	41.4	12.6	15.4	17.2	13.4
1970	44.3	18.3	9.3	14.0	14.1
1973	46.6	18.1	6.5	14.8	14.0
1976	48.3	16.2	6.4	14.5	14.5
1979	49.7	14.6	6.2	14.3	15.1
1982	50.7	13.0	6.2	13.7	16.3
1985	52.0	11.6	5.8	13.4	17.2
1988	54.0	10.8	4.9	12.7	17.6
1991	56.1	10.2	3.1	11.8	18.8

Sources: Jane Van Nimmen and Leonard C. Bruno, NASA Historical Data Book, vol. 1, NASA Resources 1958–1968, SP-4012 (Washington: NASA, 1988); NASA Office of Human Resources and Education, "The Civil Service Work Force," annual (Washington).
aCombined with clerical category prior to December 31, 1960.

The contracting ratio, which had fallen to between four and five contract employees for each NASA civil servant during the period of contraction, jumped to seven to one by 1988.[5] The proportion of the total NASA budget that went to contractors increased from 83 percent in 1978 to 88 percent in 1989.[6] NASA officials did not use increases in the space budget during the 1980s to rebuild their in-house capability.

"Now there is still a Langley Field and there still is a Lewis Laboratory, and there still is an Ames Laboratory," observed one of NASA's top executives, "but they have totally changed because they got enamored with this world of contracts."[7] Agency policy during the 1970s required NASA civil servants "to increase the involvement of outside contractors . . . and [reduce] our in-house manpower."[8] Even the research centers, where the in-house tradition was especially strong, contracted out their research activities. "NASA is a contract world," the executive complained. "It does everything contractually these days. . . . We've gone from the idea that you do it with the people you have got, to you do it with the industry that you have." This changed the demands that were placed on agency employees. "They're trying to 'manage' this world instead of 'doing' this world," he added.[9]

"The number of technical people is about the same today as it was in 1970," observed one of the program managers from the Apollo era. "But almost all of the support activities are now contracted out, so that . . . the mix of talents is different. Too much of the effort of the people in the centers is spent in managing their support contractors and not enough in doing work." The problem, he added, was especially troublesome at the research centers, where the need for in-house work was strong.[10]

Agency employees responding to the 1988 culture survey agreed that NASA had "turned over too much of its basic engineering and science work to contractors." They also sensed that NASA had lost much of its in-house technical capability as the agency matured. That sense of loss was especially strong among the people who joined the agency when it was young (see the appendix: in-house capability).

Why did the bearers of the traditional culture allow this to happen? During the first decade of operations, necessity prevailed. To mobilize the country to win the space race, NASA managers had to seek help wherever they could find it. They relied upon contractors to accomplish the Mercury, Gemini, and Apollo space flight programs because they needed help, much in keeping with the general trends that had been established in other science and technology programs during prior decades.[11] Contracting out also helped to construct a broad base of support for the new space program.

As NASA retrenched, reliance upon contractors proved more convenient. By contracting out service activities and technical work, NASA executives could protect their central corps of engineers and scientists. Scientists and engineers were the agency's dominant professional group and there was not much sympathy for pushing too many of these people out the door as the agency shrunk in size. It was easier to let trade and craft and technical support employees go and then write contracts for the same services as a way of adjusting to descending personnel ceilings. The requirements of survival were simply more powerful than the organizational culture during the period of decline. To say that NASA sacrificed part of its culture in order to save part of its work force is to admit the obvious, namely that the forces of retrenchment undermined the commitment to in-house capability.

By the 1980s, other forces worked to preclude a return to in-house capability. NASA's environment changed. At the dawn of the space age, expertise in satellites, rocketry, and space flight was confined to a fairly small group of people, many of them employed in the organizations that merged to form NASA. Corporations had little experience with space exploration, and academic institutions had yet to develop their aerospace faculties. "During this period," members of the Augustine Committee on the future of the U.S. space program observed, "it was necessary for NASA to perform a great deal of work in-house, and to provide a substantial degree of oversight to the newly formed 'space-industrial-academic complex.'"[12] By the mid-1980s, the U.S. space industry was much more mature, both in its technical capability and as a political constituency. "The arrangement which gives NASA monopoly power over decisions on the technical concepts for manned space flight," wrote a group of outside experts set up to review the Space Exploration Initiative, "has not been seriously reexamined since the establishment of the space agency more than thirty years ago. However, the circumstances which made it appropriate to give NASA this power when it was first established have changed markedly in the intervening decades."[13]

A series of presidential directives ordered government agencies to do more contracting out. Those directives dated back to the mid-1950s, when the second Hoover Commission urged government to get out of the business of business.[14] In 1966 and again in 1979, the Office of Management and Budget issued an A-76 circular directing agency heads to rely upon private industries that could provide needed services at lower costs.[15] White House officials in the Reagan administration vigorously enforced the A-76 approach. The intellectual rationale for privatization was provided by social scientists who helped promote the so-called Reagan Revolution. Social scientists presented theories of efficiency purporting to show how private firms operating in competitive environments would consistently outperform government bureaus providing monopoly services.[16]

Both Republican and Democratic administrations helped to institutionalize the ideology of contracting out in the operations of the executive branch, including NASA. As the years passed, NASA executives found themselves less able to resist White House policy. The general antigovernment philosophy of presidential executives, combined with broadening aerospace expertise, effectively precluded any significant return to in-house capability as NASA matured.

The practice of contracting out manifested itself dramatically in 1984, when NASA executives assigned overall responsibility for launch preparation of the space shuttle to a private contractor, the Lockheed Space Operations Company. Traditionally, each contractor that built a rocket component, satellite, or spacecraft sent workers to the Kennedy Space Center, where they toiled with on-site contractors and NASA personnel. Together, contract workers and NASA civil servants assembled the vehicle and its payload, checked it out, and launched it. Contract personnel performed the bulk of the work; NASA engineers integrated the work and managed the contractors. This management role placed NASA officials in a dominant position, much in keeping with the philosophy of agency oversight and contractor penetration. The first shuttle missions, including the initial test flights, were launched in this way.[17]

When NASA officials announced that they could turn the space shuttle into a fully operational transportation system, an outside consultant observed that NASA ought to be able to hire a single contractor to oversee all of the ground processing activities leading to a shuttle launch. This would include supervision of other contractors. The idea of a single contractor taking over so much of NASA's launch management responsibility was contrary to the technical culture but compatible with NASA's stated goal of making space flight more routine. The first shuttle mission processed entirely by the Lockheed Space Operations Company lifted off on April 6, 1984. Following the *Challenger* accident, when public officials had to admit that shuttle flights were not routine, NASA reviewed the processing contract. Pressures to contract out were sufficiently strong that agency executives felt obliged to retain it.[18]

In spite of the trends favoring the practice of contracting out, NASA officials clung to their original beliefs favoring in-house work and contractor penetration. NASA old-timers in particular expressed their dislike for the new emphasis on contracting out. Excessive contracting moved NASA's technical expertise out of the government and into the hands of contractors. It made contract administrators out of engineers, senior employees argued. How could the government retain the technical expertise necessary to conduct future missions if it adopted the philosophy that the government should contract out any function that industry could perform? "You have to make a balance there," said one of NASA's top space flight engineers. The balance would require NASA to spend about 10 percent of a project's budget in-house.[19]

Was NASA captured by its clientele? The word does not accurately describe the nature of the relationship. NASA executives worked hard to avoid situations like the Saturn V second stage and the development of the shuttle orbiter where they felt too dependent upon a single contractor. At the same time, by contracting out their technical capability, they ran the risk that they would become overly dependent upon contractors for the knowledge needed to run the space program. This concern was not shared by industry groups, however, which believed that NASA could retain the capability to supervise contractors even where the contractors monopolized the technical knowledge needed to manage the programs. Needless to say, this view was not shared by NASA officials steeped in the values of the original culture.

Said one NASA center director from the 1960s, "I think the success that NASA had in the Saturn program [the rocket that lifted Americans toward the Moon] was due in part to this enormously talented group that was at the Cape at the time." NASA released that talent as it matured, he observed. "One of the things they did at the Cape was to turn over the operations of the Cape to a single contractor. That should never have been done," he added. "You never hire anybody to do your work for you. You hire them to help you."[20]

"The psychology of the organization," said one of NASA's top executives from the first generation, reflecting on the changes that had affected NASA as it matured, changed to one "that says it's better to be managers than it is to be technically competent. So their measure of success was how many people you managed. Clearly, in a research organization, that's not only the wrong criteria, it's an almost disastrous criteria."[21]

Going Operational

The original NASA culture embraced the notion that space flight was inherently risky, that failure was a normal part of the learning process, that trouble was a constant companion, and that each space flight had to be treated as a discrete event. The nature of space travel during the 1960s supported this perspective, when most flights were short and daring. By the 1980s, changes had occurred. NASA officials made plans to engage in missions that varied from frequent to continuous. They encouraged the radical notion that much of what they did in space could be made safe, routine, and familiar. Such notions were totally contrary to the norms of NASA's original culture.

As organizations age, their leaders have a natural tendency to take that which is uncertain and try to make it predictable. Uncertainty is the enemy of organization. It spoils the certitude that allows institutions to survive. A business firm or government agency that is just starting up encounters many novel situations for which it has no established routines. As these institutions mature, the situations they must address become more familiar. Managers standardize procedures for dealing with those situations. Rules and schedules serve to make the outcome of work more predictable. Business firms develop elaborate models for predicting, for example, future revenues, the cost of raw materials, and demand for their products. Government agencies take on functions that must be carried out continuously, assuring a predictably steady demand for their services. Affairs that have been made predictable can more easily be managed, reducing the risk that carefully drawn plans will go awry. Without the ability to convert novel situations into familiar routines, few institutions would be able to survive for extended periods of time.[22]

This familiar tendency toward routine operations poses a special challenge for a research and development organization like NASA. The environment in which NASA employees work is inherently uncertain and plagued with risk. As people engaged in research and development, NASA employees encourage themselves to treat each mission into space as if it was a unique event requiring special attention and technical skill. They do not view themselves as conventional bureaucrats, carrying out government routines. The NASA culture supports the notion that once an operation becomes familiar, it is time to move on to something new. At the same time, the survival of a research and development organization depends upon the ability of its executives to reduce the uncertainty the organization must confront. At the least, executives would like to have a predictable source of funding. Ultimately, they must make predictable the performance of their instruments and machines.

Within NASA, the quest for predictability took the form of a shift toward space operations. This shift created a major cultural schism in NASA, a confrontation between those who saw NASA as an agency devoted to research and exploration and those who wanted to elevate flight operations. People outside of NASA frequently dwelt on the conflict between humans in space and the work that could be done by automated probes. Within NASA, however, this was not as large an issue as the conflict between research and operations. The so-called manned versus unmanned debate never produced as much anxiety within the civilian space agency as the question of how much attention NASA should devote to flight operations once the nature of flight operations started to change.

The term *space operations* has a special meaning to the engineers and scientists who work on the civilian space program. Traditionally, NASA engineers have used the term *space operations* to refer to that part of an expedition in which people and machines go into space and carry out a mission. It also refers to the training of astronauts, the assembly and checkout of machinery before launch, and the tracking and communication systems necessary to control operations once an expedition is underway. Operations as such becomes the final step in a lengthy process of research, design, development, testing, and launch. Operations so defined is a natural conclusion to the work of research and development.

During the first flights to the Moon, the operational phase for a typical human mission lasted only a few days. During the first flights to the planets with robotic probes, the operational phase lasted a few months. The longest Apollo flight to the Moon took twelve and one-half days from liftoff to splashdown. The first close-up pictures of the planet Mars were taken by *Mariner 4*, which took six and one-half months to complete its mission. Though viewed as space operations, none of these missions were treated as routine.

NASA maintains on its payroll people who are experts at space flight operations. Some work at Mission Control at the Johnson Space Center. Others work at the Deep Space Network for planetary communication on the grounds of the Jet Propulsion Laboratory; more labor at the Mission Operations Room for the Hubble Space Telescope at the Goddard Space Flight Center. Many can be found at the launch facilities of the Kennedy Space Center. They are among the best trained people that NASA employs. Traditionally, they have approached each flight with the philosophy that it is a special and unique event, subject to its own set of troubles. "The bird we are going to launch does not know that all the others flew," explained one launch executive from the Apollo era. "It stands alone, and what we do to that bird, and how we treat it, is how it is going to treat us. The fact that we had all those successes, whatever number it was, does not bear on this bird."[23]

The old philosophy of operations as the last step in a lengthy process starting with research and design was quite compatible with NASA's original culture. The new operational philosophy was not. More and more NASA civil servants found themselves preparing for operations that were, from the perspective of the early missions, essentially continuous. NASA managers deployed agency employees to work on the operation of the space shuttle, which flew frequently. They deployed people to operate the Hubble Space Telescope, which was expected to fly for at least fifteen years, and the TDRSS system, the agency's network of continuously operating communication satellites. They planned the space station *Freedom*, which, once assembled, would be continuously occupied for thirty years. They planned new additions to the operational capability of the Space Transportation System, such as the cargo-carrying *Shuttle C* and an Orbital Transfer Vehicle.[24]

NASA officials stood to gain many advantages by stressing operations. Continuous operations require continuous feeding. Congress and the White House are generally reluctant to shut down the funding for programs that are up and flying. By staking out a claim as the nation's primary provider of space operations, NASA officials guaranteed themselves a significant, predictable source of revenue each year. Predictable sources of revenue are a major source of strength for government agencies.

Even so, NASA people worried. Many were uncomfortable with the prospect of shifting so many resources to space operations in what they viewed as a research and development agency. Many continued to believe that operational activities required a different mentality than research and development.[25] Many questioned the tendency of government leaders, including some in NASA, to characterize continuous operations as essentially routine. Nowhere did these concerns produce more controversy during the 1980s than in the effort to define the status of the Space Transportation System.

NASA received approval to build the space shuttle in early 1972. It conducted the first orbital test flights in 1981. In 1982, after just four test flights of the space shuttle *Columbia,* President Ronald Reagan proclaimed that the shuttle fleet was "fully operational, ready to provide economical and routine access to space."[26] NASA executives, who provided much of the language for the President's speech, planned to launch as many as twenty-four flights of the space shuttle each year. The shuttle would become "the provider of routine roundtrip transportation to and from space," said one executive.[27] "By 1990," a leading NASA planner predicted, "people will be going on the shuttle routinely—as on an airplane."[28]

The belief that any major space flight program could be made "fully operational" or "routine" clashed with assumptions that NASA operated risky systems. "To talk about spinning that off as an operational vehicle is kind of crazy to me," said one agency executive of the space shuttle. Just because NASA operated something like the space shuttle or the Hubble Space Telescope for long periods of time did not make them routine. "Most of the things that we built were one of a kind, two of a kind, unique kinds of things," he said. "There are four shuttles, but they are all a little bit different. If you have really got something that is a production line kind of thing, then you can talk about operations in the sense that people like to talk about making something operational. But as long as you are pushing things back as far as these guys are, the shuttles having two kinds of hydrogen leaks and things like that, to me that is R&D with a heavy operations responsibility built into it." The space station *Freedom* would never become operational in the sense of being routine. "Space station will be the same way. You will continue to develop and use it for different research activities. It is just not the same in my mind as some sort of production activity. It never did make sense to me to talk about applying the same operational philosophy that the Air Force applies to their F-16s. It just never did."[29]

"NASA's culture is not one that supports an operational mode," said another top agency executive. "There has been no agency I know of that has been able to carry on effective research, effective development, and effective operational programs without separating those functions clearly and providing different management structures for each one of them." Either operations dominates to the detriment of research and development, which is typically the case, or research scientists refuse to allocate sufficient resources to operations. "If all of the technical people are working on operations, they aren't working on new technology or new anything else. They're just keeping that thing flying."[30] Another NASA executive from the Apollo era was more blunt. If NASA officials were convinced that the shuttle was fully operational, they "ought to paint the damn thing blue and send it across the river." The Air Force Department, located on the other side of the Potomac River, had more experience with operational systems than NASA did. NASA, he argued, did not have the capability to handle the logistics of a fully operational shuttle. "It's an R&D organization."[31]

Bearers of the traditional culture worried that the agency's frontier mentality would twist its approach to operations. "The first time, yes, they love it," said one of NASA's top space scientists. "But to the engineers to keep the shuttle tuned and flying for twenty years doesn't give most of them a big thrill. They will do a couple of things to you if you try to do that. They will keep thinking of new ways to improve it, whether it really needs it or not, or they will go off and do something else. If you are engineer, you want to engineer something."[32]

NASA officials close to the launch process did not accept the new operational philosophy. "The demands out there are so high," observed one launch official. "A crack in one window, you lose your crew if it blows out." The advocates of the new operational philosophy told this person that they would be able to turn the shuttle around and launch it more easily than he had launched a Saturn V. "People thought they were going to [launch] it cheaper in terms of less this, less that, less checks, and less paper. They told me they were going to launch this thing in Florida with much less people. . . . That's crazy," he said. "Look at the numbers of people they have put together to handle this next launch. There are no shortcuts."[33] Even before the *Challenger* blew

up, NASA officials had to admit that the shuttle was not "fully operational," that it still retained the characteristics of an experimental aircraft.[34] After the *Challenger* accident, the shuttle program continued to
encounter a sufficient number of difficulties to convince most people
that it would never become operational in the sense that the term is
applied "to commercial aircraft, to ships, or to mass-produced articles
of defense."[35]

Even so, the people who oversaw the space program grappled with
the question of how operational NASA might become. More and more
NASA missions, from the Hubble Space Telescope to the space station
Freedom, required continuous control. The people who embraced the
traditional culture did not want to treat those operations as safe, routine, and predictable. At the same time, they had to confront the normal
tendency of government employees to approach continuous activities
in just that way.

Flight Testing

Flight tests of experimental aircraft, rockets, and spacecraft defined
much of what NASA did during its first decade. Flight tests provided
the ultimate means of verification through which NASA officials and
their contractors discovered how something worked and made necessary design changes. Flight tests sharpened the technical skills of NASA
personnel. NASA recruited the first astronauts from flight test programs
at places like Edwards Air Force Base, where NASA maintained its own
flight test facility. The first human expeditions into space were organized
by a group of NACA engineers who had gained experience flight testing
pilotless aircraft from a remote launch facility at Wallops Island, Virginia. NASA's large rocket program was put together by engineers who
had directed guided missile flight tests at the White Sands Proving
Ground in New Mexico and the Long Range Proving Ground at Cape
Canaveral, Florida.

Proceeding from this flight test background, the first generation of
NASA space and rocket engineers spent a considerable amount of time
conducting flight tests on the machines they planned to send into space.
A review of space flight records from 1966 illustrates the importance
NASA officials placed on flight testing. NASA conducted thirty-six
launches that year. Many of the launches were flight tests. NASA officials used a Little Joe rocket launched on a suborbital trajectory to test
the launch escape system for the Apollo-Saturn vehicle that would take
Americans to the Moon. They conducted three unmanned flight tests

Table 6   NASA Flight Records: 1966 and 1985

Type of Flight	1966	1985
Launches	36 (3)	14 (0)
Launches of manned spacecraft		
With astronauts on board	5 (0)	9 (0)
Flight tests without astronauts	4 (0)	0
Launches of unmanned spacecraft		
Satellites and space probes	24 (2)	5 (0)
Flight tests	3 (1)	0
Missions and payloads arriving in space	33 (6)	36 (1)
Space flight missions with astronauts on board	5 (2)	9 (0)
Flight tests of manned spacecraft without astronauts on board	4 (0)	0
NASA satellites and space probes	17 (3)	2 (0)
DOD payloads launched by NASA	1 (0)	7 (0)
Weather, communication, and research satellites launched by NASA for U.S. government agencies and private firms	4 (1)	9 (1)
Weather, communication, and research satellites and other payloads launched by NASA for other countries	0	9 (0)
Flight tests of space hardware	2 (0)	0

Source: NASA, *Pocket Statistics,* annual (Washington).
Note: Failures in parentheses.

of the Apollo-Saturn system, two to test the rockets and one to test the spacecraft. They fired a model of the Apollo spacecraft heat shield into a suborbital trajectory to test the effects of reentering the atmosphere at high velocities. They twice tested the tricky Atlas-Centaur rocket with a model of the Surveyor space probe on board. The Centaur rocket was the first to use liquid hydrogen as a fuel; the Surveyor was a robot designed to land softly on the Moon and test whether the dusty surface was solid enough for humans to land. The five Gemini space flight missions with astronauts on board could also be considered test flights, since they involved the first attempts to rendezvous and dock with target vehicles in space, an essential flight maneuver for the expedition to the Moon.

As the space program matured, the early emphasis on space flight testing diminished. The first shift occurred during the 1960s, when the directors of Project Apollo moved to substitute ground tests for more complicated flight tests. The decision to move to all-up testing of Apollo-Saturn and flight test the entire vehicle all at once pushed much of the testing program back to the ground. As NASA matured, the opportunities for flight testing diminished even further. NASA officials test flew the space shuttle *Columbia* only four times before declaring it to be fully operational. They spent more time launching satellites and payloads for other government agencies and commerical firms.

Flight records from 1985 show how much the emphasis had changed. NASA launched the space shuttle nine times that year. Shuttle astronauts delivered twenty of the twenty-seven satellites, probes, and payloads that NASA put into space. Astronauts repaired Syncom IV-3, a Hughes Communication satellite whose sequencer had failed to start when other shuttle astronauts had deployed it four months earlier. They brought back GLOMR, a Defense Department message relay satellite, to be launched on a later flight after it failed to emerge from its canister. The European-built *Spacelab* flew on two of the shuttle flights. Mission specialists conducted materials-processing experiments and practiced construction techniques in space. These were interesting missions, but the space agency conducted no space flight tests that year.

NASA employees were generally aware of the de-emphasis on testing, especially flight tests. Surveyed in 1988, most professional employees agreed that "NASA did substantially more testing in the past than it does today." Employees from the first generation were especially conscious of this change. A majority of the entire group agreed that "NASA should do more testing than it currently does" (see the appendix: NASA testing activities).

Many blamed the decline of testing on NASA's desire to provide routine flight services. "Throughout Gemini and Apollo we were running an experimental program," said one leading executive. "When you go over to the shuttle operations and space station . . . then you're getting into an operational mode, where essentially you're running an airline and a hotel system, and you are no longer back in the business of building a different vehicle for every launch. And in that sense, the difference is great indeed."[36]

The changing technology of space flight reduced opportunities for flight tests. With the space shuttle, said one of the top Apollo generation engineers, "we didn't have a choice." The orbiter spacecraft had to be tested all-up on its first voyage into space. "We had to make the shuttle fly the first time. It was quite a gutsy thing to do. It took us an awful lot of wind tunnel tests, an awful lot of tests in the avionics laboratory to prove that it worked." With the exception of the drop tests where NASA astronauts investigated the landing characteristics of the shuttle *Enterprise*, there was no sensible way to flight test individual components of the orbiter as they had done with previous spacecraft. "It wasn't a reasonable thing to do. There wasn't any way we could break it down into a number of tests."[37]

NASA reduced its flight test activities in order to cut costs. Weaker political support and shrinking budgets forced NASA executives to cut development costs in order to get programs approved, which inevitably reduced testing. In order to reduce development costs, NASA officials decided to forgo unmanned, fully automated flight tests of the shuttle orbiter. The first orbital test of the *Columbia* shuttle took place in 1981 with astronauts on board. NASA never flew a prototype of the Hubble Space Telescope. During the first decade of space flight, NASA officials frequently used prototypes to obtain flight test results that modified the final design.

Decreased tolerance for failure discouraged testing. During the first generation, NASA officials recalled, failure on a flight test was viewed as a fairly normal event. It was the way that agency engineers learned. "There wasn't somebody keeping score in the way that you think of it now," said one NASA executive. "The philosophy was different. We knew we were learning."[38] As political support for the space program diminished and the cost of test hardware increased, failure even on a test flight unleashed a barrage of criticism. NASA officials tried to reduce the frequency of failure by conducting fewer flight tests. This did not mean that NASA cut back on testing altogether. To make up for the lack of flight tests, NASA engineers and contractors conducted ground tests. They used computer models to simulate the conditions of flight. They built in self-correcting mechanisms or maintenance capabilities that would allow them to correct failures during the operations phase.

The development of the Hubble Space Telescope illustrates the advantages and disadvantages of this new approach. In the early stages of the program, NASA officials and the panels advising them declined to recommend the construction of an intermediate-sized telescope as a precursor to the large space telescope. (NASA had launched smaller space telescopes between 1966 and 1972 with its series of Orbiting Astronomical Observatories.) They declined to recommend the construction of a prototype. Instead, NASA officials decided to use a modular approach in which components of the telescope would be tested individually by teams on the ground prior to assembly and launch, supplemented by mathematical models that replicated how the telescope would behave in orbit. They then developed plans to use the space shuttle to repair the telescope in orbit and, if necessary, return it to earth for a major overhaul.

Reduction in testing was done primarily to hold down development costs on the Hubble Space Telescope. Even though the savings from a scaled-down test program would be more than lost by the heavy costs of in-orbit repair, the political pressure to cap soaring development costs forced NASA to take the short-term perspective. It even forced them to cut back on compensating ground tests. In the original plans for the Hubble Space Telescope, NASA officials eliminated the expensive thermal vacuum tests, during which the assembled telescope would be placed inside a chamber designed to simulate conditions in space. As development problems occurred, NASA received funding to conduct additional ground tests. Many important tests were performed, including the thermal vacuum procedure, but not enough to catch a flaw in one of the mirrors.[39]

The original NASA culture viewed flight tests as the most important method for studying and verifying the performance of space-bound machines and experimental aircraft. Ground tests and computer simulations were important, but actual flight provided the ultimate test of how something worked. Routine operations, budget constraints, the diminished tolerance for failure, and the changing nature of space technology forced NASA to cut back on flight testing. Faith in the reliability of the substitutes was never as widely accepted as the original norms. NASA engineers and scientists found themselves operating under conditions that did not permit extensive flight testing even as they still held to their original beliefs about testing and verification. One more element of the original NASA culture grew weaker without a powerful substitute to take its place.

Risk and Technology

As NASA matured, the tolerance for risk and failure diminished. Failure of a satellite or piece of flight hardware was a fairly common occurrence during the first decade of exploration. As NASA moved more into operations, people expected a higher degree of reliability. Significantly, this shift in tolerance occurred during a time when spacecraft became more complicated and failure-prone.

The flight record for 1966 illustrates how frequently failure occurred in the early years (see Table 6). Of the thirty-six missions that NASA conducted that year, fully 25 percent failed, either because of launch failures or problems in space. The Atlas-Agena target vehicle for the flight of *Gemini IX* did not achieve orbit because of a short circuit in a control mechanism, and an Atlas-Centaur test flight failed when the second Centaur engine did not fire as planned. NASA lost Biosatellite I, a satellite designed to study the effects of space on living organisms, when the retro-rocket needed to bring the satellite back to earth failed. NASA lost an Orbiting Astronomical Observatory, a communications satellite launched for the COMSAT corporation, and a Surveyor probe that crashed into the Moon. Astronauts Neil Armstrong and David Scott were nearly killed when their *Gemini VIII* spacecraft spun out of control after they attempted to perform the first rendezvous and docking in space. Later that year a second docking test had to be scrubbed because of a problem with the target vehicle shroud.

Reflecting back on the Apollo era, NASA officials who served during that period expressed the belief that the agency took more risks in those days. Failure, they stated, was more acceptable then. Following two docking failures in March and June 1966, NASA astronauts John Young and Michael Collins successfully docked with their Agena target vehicle in July. NASA followed up the failed *Surveyor II* mission with the successful landing of *Surveyor III* the following year. After *Surveyor IV* disappeared three months later, NASA officials successfully landed *Surveyor V* on the Moon. Space probes went out in series, so that a failure on one mission did not doom the whole program. Failure on one became the basis for improving the next. The public, accustomed to seeing rockets explode on their launch or test stands, did not expect perfection. NASA officials brought all of their astronauts back alive, not because they assumed perfection, but because they assumed that failures would occur.

NASA achieved an outstanding reliability record in 1985. Of the thirty-six missions and payloads carried into space that year, only one failed—a satellite launched from the space shuttle *Discovery* for Hughes Communication. All fourteen launches, including nine of the space shuttle, were successful. This apparently improved flight record masked an important transformation in the nature of the U.S. space flight business. All but two of the payloads launched into space that year were delivered by NASA in its role as a provider of routine earth-to-orbit transportation services. NASA did not conduct a single test flight of a space vehicle that year. People expected NASA, as a provider of reliable flight services, not to fail.

As much as they lamented the passing of the first era, NASA's engineers and scientists recognized the limitations that a new generation of spacecraft and space probes placed on their ability to endure failure. "It is so much more expensive now," said a top NASA executive. "The payloads that we fly are so expensive. The problems that we have are expensive both in costs and the time it takes to turn the program around and get going again. So I think we have to reduce the risks that we have."[40]

An official from the space science side agreed. "You know, when you invest the resources in a planetary mission that takes three to five years to conduct after the launch, you are not as tolerant politically during the science [phase] of being able to accept a failure."[41] Long-term missions clearly affected NASA's outlook toward failure. "The Hubble Space Telescope is going to introduce a new series of vehicles which are intended for very long periods of life," the executive added. "If the Hubble Space Telescope ceases to operate in a way that we can't repair it, then that is going to shut down a whole field of endeavor for many, many, many years. We were never that dependent upon an individual scientific instrument before."[42]

Not only did flight systems grow more expensive, they also became more complicated. To use a favored term, space technology became *tightly coupled.*[43] Using tightly coupled systems like the space shuttle, NASA was able to improve the performance of its hardware. Tightly coupled systems, however, affected the nature of the risks that NASA faced. One of NASA's top Apollo engineers explained.

"If you try to, you know, use all of the pig except the squeal, so to speak, you're obviously making things work very efficiently. But, in order to do that, you end up with a more complicated system. A case in point is the space station. You could interrelate the propulsion system and the life support system in several very interesting ways." Life support requires oxygen. Propulsion requires oxygen and hydrogen. "Well, the thought comes right away, why don't you bring water up and generate oxygen and hydrogen." It is much easier to carry water into space than to transport liquid oxygen and hydrogen. "So, you interrelate, you intertwine the two systems together because both have to make oxygen. The only problem is that when you start doing this, failure in one of these things cascades over in the other."

The same principle applies to rocket design. By separating the components that make up a rocket, the likelihood of catastrophic failure can be reduced. "At some cost of performance," the engineer continued. "You're letting some opportunities for increasing your performance get away when you do that. I'm basically inclined to try and make things simple as opposed to high performance every time, although I've been dealing with high performance systems all my life."

Recent trends, he admitted, had not carried NASA in that direction. "The trend is to greater complexity in order to get more bang for the buck. . . . In a case like the shuttle, if you don't take advantage of some of the opportunities to improve the performance, you'll never get off the ground." In a simpler design, the three main engines could be taken off the shuttle orbiter and placed on a separate launch vehicle. That would reduce the complexity of the orbiter. It would also reduce the efficiency of the space shuttle by making it harder to recover the main engines. "One of the things about the shuttle that makes it efficient is that we take part of the propulsion system—the launch propulsion system—and put it in the orbiter during launch. All the main engines, great big ducts, all that ends up in the rear end of the orbiter." The main engines thus come back to earth when the orbiter lands. "Well," the engineer concluded, "that is a big complication."[44]

During the first generation of space flight, as one of NASA's launch directors observed, launch vehicles had a simpler design. "Things were simpler when it was a one-stage rocket with one engine and a capsule. Your probability of having a problem was less because there were less components and your probability of catching it before it became catastrophic was much higher." Problems were easier to find.

Launch vehicles became more complex as NASA progressed from Project Mercury to the space shuttle program. "Now a sinister little thing can happen down here in some component that is two or three levels removed from the visibility you have when you are doing the checkout," the launch director continued. "You hope that as you have developed a system, you give yourself enough measurement points and you certify it and you do a lot of testing on margin and on failures to catch these things. But the bigger the systems get and the more interactive and the more interfaces, the higher the probability that you won't catch it all."[45] NASA professional employees agreed that "the possibility of an interactive failure on a manned or unmanned space flight has increased" as the technologies they used became more complex (see the appendix: interactive failures).

The interactive complexity of new space flight systems together with the length of time they operated in space changed the way that NASA employees managed risk. The Apollo astronauts on ten flights to the Moon (of which six landed) spent a cumulative total of 92 days away from Earth. NASA had one major accident on *Apollo 13*. Estimates for a Mars expedition ranged from 500 to 1,000 consecutive days away from Earth.[46] NASA expected to operate the space station in low earth orbit for up to thirty years.

"It is putting a demand on our abilities to accommodate failures," said one top executive, "which is different than the demand that we had when we were throwing everything away, when everything was expendable." When missions were short and launch systems retired after every flight, NASA engineers could tolerate a component that, given statistical probabilities, had a relatively short expected life. "We could live with that." On long-duration activities like the space station or a Mars expedition, this approach will not work. On such missions, equipment will fail. It is inevitable. "We have to have the capability to sustain failures and work our way through those failures without catastrophic results." This will require built-in, on-board, diagnostic capabilities—built into the spacecraft and not on the ground. It will require in-flight maintenance, component replacement, redundancies, and "a variety of techniques, some of which we haven't yet invented."[47] The engineering challenge will be enormous.

As an engineering organization, NASA's basic approach to risk and failure in complex systems has been to design its way through them. Whether by designing machinery that lasts for the duration of a mission, or by emphasizing replacement and redundancy, NASA engineers seek to "build up confidence by design."[48] Some NASA officials wondered whether this actually could be done. "I don't believe it," said one of the agency's top scientists. "As spacecraft become more complex, the interactions from one part of the system to another become more complex and you have more potential failure points. You try to account for some of that by building in redundancy, alternate pads, cross strapping from subsystem to subsystem, but complexity is getting to be such that you can't build in enough redundancy to cover all the possible failure points, nor can you test the number of ways things can go wrong."

Complex systems work in unexpected ways. "Fifteen years ago, if you had failure X you could rapidly pin it down to the part of the space craft where the failure really occurred and work your way around it, or try to work your way around it." The way that failures occur in modern spacecraft, the scientist observed, "the damn things could have really happened in about five different ways or ten different ways." The interactions are such that a fix to one part of a spacecraft may affect the performance of another part. "The engineers are trying to come up with design approaches," the space scientist concluded, "but it's not going to be the total solution."[49]

These developments took place at a time when the tolerance for failure on NASA space flights approached zero. Although agency employees continued to believe that "risk and failure are a normal part of the business of developing new technologies" (see the appendix: risk and failure), the public discussion of risk by NASA employees was practically taboo. NASA officials refused at first to even conduct formal risk analysis on the space shuttle and then issued qualitative judgments that set the probability for a catastrophic failure as low as one in 100,000. As physicist Richard Feynman pointed out, such a probability "would imply that one could put a shuttle up each day for 300 years expecting to lose only one."[50]

NASA officials apparently believed that a frank discussion of risks would doom agency projects for which political support was already fragile. The political system demanded risk taking but discouraged failure. These lessons had been ground into the space program since the mid-1960s, through incidents large and small. The Ranger probes, the Apollo fire, the *Challenger* accident, and the difficulties in getting a long series of new projects approved all reinforced this perception. It persisted even as NASA officials moved toward new approaches for managing risk on long-duration missions. NASA's capacity to conquer new frontiers by taking risks and enduring occasional failures diminished as the space program matured.

The Distance Thesis and Exceptional Employees

As NASA matured, professional employees continued to believe in the exceptional quality of the agency work force. By a margin of nearly two to one, NASA employees responding to the 1988 culture survey agreed that "NASA still recruits exceptional people." Although the first group of NASA employees believed that their generation was "truly exceptional," they gave high marks to new recruits as well (see the appendix: recruiting exceptional people). A 1991 study by the National Academy of Public Administration concluded that "the quality of new hires today is as good or better than in earlier periods."[51] Survey data on educational backgrounds showed no significant diminution in the qualifications of new recruits (see the appendix: educational backgrounds). Even outside experts interviewed for this study agreed that NASA still recruited exceptional people, at least by government standards. The only persons who disagreed were those who worked for top-level academic or research institutions, where recruiting standards were unusually high.

"We're still recruiting exceptional people," said one of NASA's top engineers. "What you're seeing now is a new generation of people. You know, space was very attractive to the young people when we were doing Apollo. The ones who were in kindergarten and elementary school are now showing up as your recruits. They got marked with the program at that time, just like I built model airplanes and became a space nut. They want to be part of it, a childhood ambition. Very exceptional people." The ones who wanted to work in the space program had to complete challenging courses of study. "As you might imagine," the engineer continued, "it was only the brighter children that could really understand all of the technical aspects of space when they read about it in books and all. That was the leading technology program in the world and they wanted to be part of it, so they got nailed very early." NASA, he concluded, could still attract exceptional people.[52]

Although NASA might still recruit exceptional people, the new recruits encountered a considerably changed work environment in the civilian space agency. The agency had grown more bureaucratic, more preoccupied with administrative activities, and more oriented toward routine. Much of NASA's technical work had been turned over to contractors, opportunities to conduct flight test programs had diminished, and the tolerance for risk and failure had decreased. The original NASA culture placed a premium on the development of an exceptional work force with strong technical skills; the new culture placed more emphasis on administrative talent. New recruits found themselves presented with fewer opportunities for sharpening their skills through the type of hands-on work available to the first generation of NASA employees. The distance between the average NASA professional employee and the technical work of space flight increased as the agency matured.

By the late 1980s, just 3 percent of the professional work force in NASA reported that they spent most of their work week "working in a laboratory, test facility, control or tracking center, training astronauts, or working on space flight or aeronautics hardware." What did the rest do instead? Seventy-seven percent of NASA's professional work force—primarily scientists and engineers—reported that they spend most of their time just like any other government employee, "working at a desk in an office" (see the appendix: work of the NASA employee). Some did analytic work at their desks using computer programs, a technical task. Working with computers, however, did not produce the same sort of hands-on experience as working with machines. Much hands-on work passed to aerospace firms, and 77 percent of the NASA employees surveyed reported that they spend at least some of their work week supervising contractors.

"All of those people who became the NASA of the early 1960s had a lot of experience designing, building, flying things and all," said one space scientist from the Apollo era. The people from the second generation of space flight, he added, did not get the same kind of flight experience or opportunity to design and build and fail. "It has been twenty years since the space shuttle program started, and a lot of people have been involved in only one program. And how have they been involved? Well, not by doing research and flying things or building them and testing them. It's by monitoring contracts." There was no doubt in his mind that NASA had lost a great deal of the technical skill created by the hands-on experience of the early days. "I can't prove it, but I think not having that kind of capability will change the capability of NASA."[53]

NASA employees certainly missed the opportunities afforded by hands-on work. Responding to the 1988 culture survey, nearly two-thirds of NASA employees reported that they did not "have as much opportunity to do 'hands-on' work as we want." Of the cohort who joined NASA prior to 1970, 84 percent reported that when they first came to the agency, "we did a lot more 'hands on' work than we do today" (see the appendix: hands-on work).[54]

How did this affect NASA? The growing distance between NASA employees and the actual work of space flight weakened the agency's technical culture. It reduced opportunities for NASA employees to practice their technical skills once they joined the agency. This occurred as a new generation of space activities—from the great observatories to a possible journey to Mars—imposed sterner requirements for technical capability on the average NASA employee.

NASA's strength as a government agency has always rested with "the quality of its technical people," said one of the agency's top executives. "That's it in a nutshell." NASA employees understood the objectives of the agency, it was easy to tell whether they succeeded or failed, and they took their technical work very seriously. "They were almost single-mindedly dedicated to quality, doing the job right, and they worked [very hard] understanding the programs in depth. They asked good, penetrating questions and insisted on quality answers." A person who came to a meeting unprepared got chewed out, "and so you didn't do it very often."

"That has gone down hill," he explained. "On the one hand, the upper levels of management have been diverted by the Washington bureaucracy into having to spend a hell of a lot more time on nontechnical subjects." Managers were no less qualified, he observed, but the demands of the job required them to spend a great deal of time on administrative and political affairs. At the other end of the organization, "Our whole attitude in demanding technical and cost accountability has slackened." The quality of technical discussions within the agency had diminished, the executive observed. "Our meetings are not as open as I think they used to be in the sense of having good technical arguments and penetrating questions asked. I've been at too many meetings where somebody will start out on that tack and the recipient gets extremely defensive." They take it personally, as a threat to their program rather than as a chance to debate technology. "It's 'What the hell has he got against us or our program?'" the executive said, reporting what he had heard others say.[55]

Seeing these trends under way, people inside and outside the agency worried about the future of the space program. They worried about NASA's capability to recruit and retain exceptional people. Although NASA employees believed they still recruited exceptional people, they felt insecure about their ability to do this in the future. Much of that concern was directed at the civil service system.[56] The rules and regulations, the pay gap, the conflict of interest laws, and the difficulties of removing nonperformers all seemed to work against the maintenance of a highly qualified work force. Young people came into the agency full of competence and enthusiasm, and the system wore them down. It pushed them back toward the bureaucratic norm. As bureaucracy increased, and the opportunity to do technical work decreased, the motivation binding exceptional employees to the civilian space agency diminished. While most NASA employees agreed that "working on the space program is still just as exciting as I thought it would be," a substantial number did not (see the appendix: the lure of space exploration). Without the motivation provided by the work itself, given the limitations of the civil service system, NASA could lose its ability to attract talented employees.

Organizational culture played a special role within NASA. It helped to motivate high performance. By assuming themselves to be exceptional people, NASA engineers and scientists could justify the practices that allowed them to perform at very high levels. They could justify their desire to maintain close control over the work of contractors. They could demand exceptional performance from their colleagues. They believed in their ability to take risks and come through safely. They defended their hands-on work, both because they were good at it and because it was a method for maintaining technical skill. The faith in exceptional people created a foundation that served as a base for many of the tenets of NASA's original culture.

Conclusion:
Governmental Performance
and Cultural Instability

*We choose to go to the moon in this decade and to do
the other things not because they are easy, but because
they are hard.*—John F. Kennedy, 1962

Congress created the National Aeronautics and Space Administration
in 1958 for "the preservation of the role of the United States as a leader
in aeronautical and space science and technology."[1] In 1961 President
John F. Kennedy gave NASA the long-range goal "before this decade
is out, of landing a man on the moon and returning him safely to
earth."[2] Borrowing extensively from the agencies that came together to
form NASA, agency employees fashioned a strong organizational culture. That culture supported exceptionally high levels of performance
for tasks very difficult to perform.

The culture was not long-lived. Even before Project Apollo drew
to a close, the culture began a process of transformation. Practices that
had exemplified the early culture gave way to different organizational
norms. Some of the original beliefs changed. NASA took on the characteristics of a more conventional government bureaucracy.

What was it that made the original culture work so well? Why
did that culture change? The NASA experience suggests that high-
performance cultures of the sort initially developed within the civilian
space program are inherently unstable. Forces that affect the government as a whole undermine them. This chapter describes the similarities
between NASA's original norms and the cultures found in other high-
performance organizations, and suggests why these cultures invariably
wither when tried in modern government.

159

NASA's Original Culture

The people who created NASA assembled a confederation of cultures, a combination of traditions from existing government agencies each with its own philosophy of doing things. NASA's early cultural norms emerged from common traditions and clashing beliefs, shaped by the requirements necessary to get the space program under way. Interviews with people who worked for NASA during these years, supported by survey data and material from NASA archives, systematically reveal those norms.

From the beginning, officials representing the organizations that came together to form NASA advocated research, testing, and verification as the primary methodology for determining what worked and what did not. Agency employees wanted to test their hardware exhaustively, even the products that came ready-made from contractors. They accepted testing and verification, especially flight tests, as the primary means for determining truth.

In order to carry out their work, officials from the predecessor agencies firmly believed that NASA had to maintain its own in-house technical capability. Shops and laboratories on the grounds of the field centers where NASA employees worked physically supported that capability. In-house capability provided a training ground for space age managers by requiring engineers and scientists to take responsibility for entire projects. It provided NASA employees with the know-how needed to effectively monitor the work of contractors.

NASA employees wanted America's civilian space agency to be a hands-on organization. From their government laboratories to Mission Control, NASA employees wanted to work with the machinery, do space science themselves, and closely supervise the work of contractors when contractors became involved. Hands-on work kept NASA employees technically sharp, and it helped agency executives attract exceptional people to the civil service work force.

The assumption that NASA attracted exceptional people was a key element in the agency's culture. Since they believed that they had recruited exceptional people, agency leaders felt justified in adopting procedures that exceptional people could carry out. The expectation of hard work and the agency's much-touted attention to detail drew on the assumption that NASA had recruited people who would behave in exceptional ways. So did the expectation that NASA employees possessed the know-how to tell contractors exactly what the agency wanted done.

The original technical culture contained strong norms about risk and failure. As members of a research and exploration agency, NASA employees viewed risk and failure as an inevitable part of their business. They expected their people to take risks and they tolerated mistakes, although not the same mistake twice. They viewed failure as the normal way by which engineers learned to improve the rockets and spacecraft that they flew.[3] NASA employees expected trouble to occur. Because they anticipated trouble, they felt better prepared to avoid it. They developed techniques for lowering the likelihood of failure, including communication networks, redundancy engineering, and extensive testing.

The people who staffed the new space agency shared a frontier mentality. They did not want to become just another government operation performing the same tasks year after year. They wanted to design and test and build and move on to new challenges. They became advocates for new initiatives. They developed new technologies where necessary but relied upon proven technologies where possible.

The creation of this technical culture during NASA's early years received support from the values that the first generation of NASA employees and scientists brought to the infant space program. Raised during the Great Depression and the Second World War, this generation of employees accepted the middle-class values of honesty and hard work as natural parts of life. As first-generation entrants to a professional corps, they pursued technical solutions to space flight problems not only because they were trained to do so, but because engineering and science had provided their means for climbing the social and economic ladder.

Together, these elements created a culture in which NASA civil servants felt empowered to exercise a high degree of discretion and technical judgment in carrying out their work. Such characteristics are typical of high-performance organizations. In her study of high-performing firms in the private sector, Rosabeth Moss Kanter suggests that "integrative" companies develop a greater capacity for innovation than firms that adopt more traditional or "segmentalist" norms. Integrative firms rely on a decentralized matrix structure rather than a centralized hierarchy. Information flow is open and horizontal in integrative firms. Their organizational culture is clear and tends to favor individual initiative and risk taking.[4]

In their study of public agencies that perform operational tasks at high levels of reliability, Todd LaPorte and other members of the High Reliability Organization Project also emphasize the importance of local responsibility. LaPorte and Paula Consolini describe how strict military authority on a navy aircraft carrier gives way to local discretion as jet traffic increases or weather deteriorates: "Collegial authority (and decision) patterns overlay bureaucratic ones as the tempo of operations increases. Formal rank and status declines as a reason for obedience. Hierarchical rank defers to the technical expertise often held by those of lower formal rank."[5]

Workers in high-reliability organizations restructure themselves into new formations as the tempo of work picks up or unanticipated situations occur. According to LaPorte, the hierarchy that normally guides work when affairs are slow steps aside so as to allow employees on the action line to exercise more discretion. Agency chiefs insist upon frequent communications in order to anticipate and correct errors. While NASA managers face different challenges than those confronting aircraft carrier captains at sea, both rely upon cultures that stress open communications, technical judgment of people in the field, and—most important—an overlay of strong control.

Observers should not be misled into thinking that the early NASA culture emphasized only technical discretion. To complete complicated programs like Project Apollo, NASA executives installed tight management systems. They imported rigorous management practices from the Air Force Ballistic Missile Program. George Low, manager of the Apollo spacecraft and later Deputy Administrator of NASA, explained the importance of the latter. "We established design standards which all of our systems had to meet and developed rigid procedures to assure that they were met," Low said. "We placed all changes under the most rigid of controls. Emphasis was on formality and discipline at every step along the way. Manufacturing and assembly were also carried out to exacting standards. . . . We specified how to solder, how to crimp wires, and controlled the process of plumbing. Every part of the system was known, its manufacture specified, and the people who performed intricate functions were specially tested and certified. . . . Formality, discipline and rigor were the key words in the test program. Test specifications were prescribed in advance, test results were audited and certified, all anomalies were reported, and all failures had to be understood and corrective action taken."[6]

Throughout the first decade, in large programs like Apollo, organizational pressures of program management and underlying norms of technical discretion coexisted. The different requirements of each produced a state of tension. The tension manifested itself in controversies over the proper role of contractors, testing philosophy, and the degree to which officials at the NASA field centers would have to bend to a central office. Frequent conflict and a norm of wide-open communications accompanied this tension, allowing employees to sort out their differences and reach decisions. The tension between organizational pressures and technical requirements motivated NASA's early culture. It is characteristic of other high-performance organizations as well.

To energize the Manhattan Project that created the atom bomb, the U.S. government combined powerful forces of organization with a strong technical culture. Scientists under J. Robert Oppenheimer provided the technical base. Organizational pressures were created by the U.S. Army under the project commander, Brigadier General Leslie R. Groves.[7] High-reliability organizations combine tight management systems with local responsibility. The official hierarchy in such organizations can be quite strict. Leaders stress the importance of preprogrammed decisions, inculcating standard operating procedures into the work of operators through rigorous training. The tension that results is similar to the "loose-tight" properties observed by Peters and Waterman in their study of highly successful business firms.[8] It is also analogous to the properties found in "dialectical" organizations developed to deliver social services to low-income families.[9]

Organizational Culture and Change

The balance between strong technical discretion and tight management control did not last long in NASA. The second generation of NASA engineers and scientists did not receive the same latitude for technical discretion that the original members of the space agency were allowed. The second generation inherited an organization with much weaker central control and far more bureaucracy.

As NASA matured, it lost much of the administrative flexibility that characterized the early years. Bureaucratic procedures proliferated. Even as NASA employees held fast to their faith in the value of risk taking, they sensed their leadership becoming more cautious and less inclined to tolerate failure. Officials became more concerned with the survival of the organization. They adopted management schemes that sacrificed institutional flexibility in order to ensure institutional survival. The tight management practices imported from the U.S. Air Force disappeared. In their place, NASA executives adopted a weaker lead center system of decentralized responsibility. Overall, a subculture of institutionalized bureaucracy took hold.

As central controls grew weaker, so did the old technical culture that once counterbalanced them. Increased pressures to contract out research and exploration activities diminished the tradition of doing sufficient work in-house. The desire to convert aspects of the exploration program such as the Space Transportation System into repetitively, predictable operations undercut the belief in extraterrestrial flight as inherently risky and bound to change. NASA officials did less flight testing and advanced the notion that computer analysis and ground testing provided an adequate substitute for the old norms. Professional employees watched the tolerance for risk and failure decrease at a time when their space vehicles and platforms grew more complex and prone to breakdowns. Even as they continued to believe in the exceptional quality of the NASA work force, they saw the distance between the average NASA employee and the business of space flight increase, diminishing the hands-on orientation that had been such an attractive feature of the original space program.

In the business world, organizational cultures do not easily change. Employees hold tight to the tenets of their cultures, resisting developments that would alter underlying norms. For many business executives, stable cultures create obstacles to needed change. Following the breakup of the Bell System, for example, managers at the American Telephone and Telegraph Company (AT&T) sought to change the underlying beliefs of employees so as to make the company more competitive. Among other goals, they sought to replace the employees' sense of technical superiority with a desire to compete openly in consumer markets. The employees responded by taking their beliefs underground, holding on to them in private and waiting for an opportunity to restore them to their rightful place.[10] In his discussion of cultural change, Edgar Schein notes the difficulty of changing fundamental assumptions and beliefs. As he observes, "Considerable change can take place in an organization's operations without the cultural paradigm's changing at all."[11]

NASA faced the opposite problem as it matured. Its high-performance culture underwent a process of transformation starting when the agency was barely ten years old. Organizational cultures consist—at a minimum—of practices, beliefs, and assumptions. Practices in NASA changed considerably. Beliefs changed some. Assumptions changed hardly at all. The overall culture changed in the sense that the remaining beliefs and assumptions lost their power to elicit appropriate practices. NASA professional employees, for example, held fast to their beliefs about risk and failure, but those beliefs no longer produced the desire to actually take risks and possibly fail. Results from the NASA culture survey confirmed the extent to which the second generation of NASA employees continued to hold the same beliefs as the first. On seven key questions, NASA employees registered their agreement with traditional beliefs about testing, in-house technical capability, hands-on work, exceptional employees, risk and failure, and change (see the appendix: questions 2, 3, 5, 7, 8, 18, and 29). In every case save one, the beliefs of employees who joined NASA after 1969 did not diverge significantly from the beliefs of employees who joined in 1969 and earlier. Only in their desire for hands-on work did the newer generation slip away to any significant degree from the beliefs of the first generation.

The original generation of NASA employees passed on to the second generation many beliefs from the original era. NASA recruited people into the agency who—for whatever reason—continued to profess the original tenets. They found themselves incapable, however, of translating those basic beliefs into the full-blown culture of the first generation.

If the original culture was so successful, and its beliefs so pervasively held, then why did it change? Three factors contributed to the breakup of NASA's original culture. The environment supporting the early space program changed. The government as a whole grew more bureaucratic. And NASA aged. These developments created a basic instability between the assumptions and beliefs of NASA's early culture and the practices needed to support them. People in NASA had little control over these developments. The inevitable transformation of NASA shows how difficult the maintenance of a government culture supporting high-performance standards can be.

Cultural Environment

Invariably, the political environment that gives rise to a new agency creates favorable conditions for its development. Political environments, however, are quite volatile. Favorable conditions quickly disappear. It is very difficult to reproduce favorable environments once a government agency matures.

NASA appeared on the government scene at a time of extreme cold war tension during which the Soviet Union seriously challenged America's sense of technological superiority. If the Soviet Union could beat the United States in a space race, could it not also pull ahead in the quest for military superiority? NASA became a principal mechanism for affirming the supremacy of American technological know-how and its spirit of can-do. The space race during its infancy had an emotional value that went far beyond the scientific or technological value of the missions themselves.

To meet this cold war challenge, NASA engineers and scientists were given enormous technical latitude. Congress and the White House gave them a great deal of money.[12] They gave them a precise mission, generally viewed at the time as nearly impossible to accomplish, and a short period of time within which to do it. The freedom to exercise technical judgment, to take risks, and even to fail is enhanced in government under conditions such as these.

Project Apollo, the race to the Moon, was set up as a crash program. The strong sense of mission that accompanies a crash program provides government executives with their most reliable method for motivating agency employees to perform at very high levels. Crash programs help to clarify cultural norms as organizational beliefs are tested against pressing realities. Crash programs, however, always end. When they do, the agency that remains must confront the problem of maintaining its culture and its high level of performance under not so favorable conditions.

Changing environments alter governmental cultures. The NASA experience with flight testing illustrates how this is so. Between 1958 and 1965, NASA's budget steadily increased. At the same time, with space missions in a seemingly primitive stage, the cost of many rockets and spacecraft remained relatively low. Low costs and expanding budgets provided NASA with the opportunity to produce additional flight articles for testing. During Project Apollo, NASA and its contractors produced thirty-three command and service modules and twenty lunar landing modules at a per-unit production cost of approximately $55 million. Less than half of them carried astronauts into space. Twenty-five of them were used for ground and flight tests.[13] With more rockets or spacecraft on hand, it was easier for NASA managers to sustain a failure on a single test flight so long as no humans were on board.

By 1970, the political environment of the American space program had changed considerably. The contraction of NASA's budget and work force was well under way. At the same time, the real cost of most missions increased, even as agency officials sought mightily to reduce costs with programs like the Space Transportation System. Fewer dollars bought even fewer spacecraft. During the early years, between 1960 and 1972, NASA programmed $355 million for the Orbiting Astronomical Observatory program, which bought four satellites.[14] The

Ranger program, which produced nine spacecraft, cost a total of $259 million in programmed funds between 1960 and 1966.[15] By contrast, NASA spent over $1.6 billion between 1978 and 1986 to design and build a single Hubble Space Telescope.[16] Contrary to the experience in business, where mass production and competition tend to reduce per-unit costs, the cost of NASA programs increased with time as their complexity grew, even adjusting for the value of the dollar.

Larger costs and smaller budgets reduced the opportunity for flight testing. NASA could not afford to request a few extra flight articles for testing when politicians choked on the cost of the main article. NASA officials turned more toward ground tests, where problems could be fixed before the main article flew. Aborted ground tests, moreover, engendered fewer political groans than flight test failures, especially for high-dollar items. Ground tests, however, could not duplicate flight conditions exactly, which increased the odds of failure in flight. Under the conditions of its new political environment, nonetheless, NASA officials had to learn to live with more ground tests.

The maturing of the American space program also affected NASA's commitment to in-house work. Between 1958 and 1965, as NASA built up the size of its work force, it assembled behind its doors many of the most knowledgeable people in the world in the technology of space flight. At the same time, because space flight was new, few competing centers of talent existed outside of NASA. As the Augustine committee later explained, "external space expertise hardly existed, and an aerospace industry had yet to be developed from airframe companies, their suppliers, and a nascent electronics industry."[17] NASA possessed the capability to do a great deal of its own work and to tightly monitor the work of contractors. NASA Administrator Keith Glennan, who favored a policy of extensive contracting out, nonetheless observed in 1960 that "most of our center directors believe that they can do the job faster, better and more cheaply than would be the case if they employ industry. In some instances, I think they are right."[18]

As NASA matured, so did the aerospace industry—in no small part due to the efforts of Glennan and James Webb, NASA's second administrator, to build it up. When bearers of the old cultural norms sought to restore NASA's in-house capacity with funds appropriated in the 1980s for new programs like the space station *Freedom*, they encountered a much more robust aerospace industry.[19] "The environment has changed," argued members of the Augustine Committee. "There is now a large and experienced space-academic community and an industrial base whose skills are broad and deep." The Augustine Committee called upon NASA to adopt a more realistic approach toward in-house capability. "The Department of Defense sponsored National Security Space Program is almost twice the size of NASA's program," the committee observed, "but operates with only limited in-house laboratory support."[20]

In its early years, NASA could insulate itself from political pressures that might force cultural change. As it matured, that insulation disappeared. NASA became more accountable to Congress, the White House, and other groups with which it had to deal. This is a natural development in a government that requires multiple accountability among its public executives.[21]

The demands for increased accountability weakened the ability of agency employees to hold on to beliefs that did not conform to objective reality. Groups that can insulate themselves from outside pressure have a much better chance of maintaining such beliefs than those that are not so situated.[22] NASA employees knew that risk and failure were constant companions on space flights. Politicians and members of the general public, who had seen the frequency with which rockets blew up, seemed to know this too. NASA officials of the first generation used this knowledge to strengthen their belief in failure as a normal part of their job.

This belief, while widely held by NASA professional employees, was not objectively true. Politicians do not tolerate failure. Opponents of individual programs use catastrophes as devices for forcing a reconsideration of priorities. Even friends of programs use accidents as a means for strengthening political control. NASA seems to have insulated itself from this reality during its early years. The high priority granted the space program, respect for the assassinated president who had helped to launch it, and the luck of not having a really serious accident until 1967 all worked in NASA's behalf. NASA officials seemed genuinely surprised by the lack of tolerance for catastrophic failure when it actually occurred, following the fire in 1967 on Launch Pad 34 that took three astronauts' lives. The reaction to the *Challenger* explosion in 1986 was even more severe. The tolerance for really serious errors never seems to have been high—either outside NASA or in the upper levels of the NASA hierarchy—a fact adequately demonstrated by the failure of the first six Ranger probes.[23] It took many years for this to sink in, and it never displaced the underlying belief inside NASA during the first three decades of space flight that risk and failures could be excused as a normal consequence of agency responsibilities.

During the period that NASA's ability to resist demands for accountability wore thin, the number of demands increased. The administrative burden imposed on all federal agencies grew. Congressional staff tripled in size in the thirty years following the dawn of the space age. NASA executives who could once resolve problems by dealing with a few well-placed congressional leaders found themselves responding to multitudinous staff inquiries. The White House policy review process became more complicated. NASA executives who had earlier dealt with high-ranking White House officials, including the President and Vice-President, found themselves negotiating with White House minions whose approval was required before moving on. Civil service regulations proliferated. Procurement regulations metastasized.

These changes reduced NASA's technical discretion and accelerated the agency's slide toward a more bureaucratic mode of operations. Engineers and scientists who once had a relatively free hand in making what they viewed as technical decisions had to clear their work with congressional and presidential aides. The choice of lunar orbit rendezvous in 1962 as the means for taking Americans to the Moon was made wholly within NASA, even though the President's Science Advisor tried to get involved. By contrast, NASA engineers had to share the design of the space station *Freedom* with a variety of presidential aides, lawmakers, congressional staffers, and outside advisory groups for a seven-year period beginning in 1984. Not only were outsiders less willing to defer to NASA's technical judgment, there were more of them pressing their oversight claims on the agency.

NASA managed during its formative years to reduce the weight of bureaucracy by rotating people in and out of the agency, To build a bureaucracy, government officials must establish a career civil service: a group of lifetime employees whose material and social condition depends upon their willingness to carry out agency policy. Members of professions resist this. They do not like to tie their livelihood to a government job. Those who can move freely between jobs give public agencies a large measure of flexibility and freedom from bureaucratic control. To raise professional loyalties over bureaucratic allegience, however, these people need the freedom to move. Easy rotation in office thus provides one of the principal means for counteracting bureaucracy.[24] Within NASA, rotation in office also supported the assumption that the agency could (and would) recruit the very best people into the American space program. Through its open door policy, NASA could bring them in from wherever they were, in part by promising them that they did not have to stay too long.

As NASA matured, Congress passed laws that made this harder to do—not just in NASA, but in all government agencies. In 1978 Congress passed the Ethics in Government Act, seeking to restrict abuses arising from the so-called revolving door. Specifically, the law prohibited a former government employee who took a job in the private sector, for a period of two years, from working on any government project in which he or she had been involved as a civil servant. In 1989 Congress amended the U.S. Code to permanently prohibit a former government employee from working on a government contract in which he or she had been previously involved, for the entire life of the contract.[25] This had a chilling effect on NASA's tradition of lateral movement among government, industry, and academia. It required scientists and engineers outside NASA to face the possibility that they might have to give up their professional interests for at least two years as a price for taking a tour of duty with the civilian space agency. Alternatively, they had to contemplate a lifetime commitment to a government job.

The Ethics in Government Act had a noble objective, namely a reduction in influence peddling by former government employees. Within the American space program, however, it had another distinct effect. Fewer employees rotated into NASA from outside the agency (see the appendix: working for industry). The act took its place alongside many pieces of administrative legislation that Congress passed during that period, one more general administrative regulation restricting the flexibility of NASA scientists and engineers to carry out the beliefs contained in their organizational culture.

Organizational Aging

While many forces impeding NASA's organizational culture arose outside the agency, others appeared from within. This is not to say, however, that NASA officials had any more control over them. If life cycle theorists are correct, then internal forces arise as a natural result of an agency growing old.

Life cycle theorists predict that management procedures will grow more elaborate and formalistic as an institution ages. They predict that the goals of the organization will shift toward those that maintain its survival, and that second-generation employees will lose their capacity to resist tendencies such as these. Faith in the ability to beat the system and avoid formal procedures was much reduced in the new NASA, as responses to the culture survey revealed (see the appendix: the NASA bureaucracy). Management systems became more elaborate, but not because NASA employees wanted to make them so. In fact, NASA officials sought to reduce the burden of formal management control.

During Project Apollo, NASA executives developed strong management procedures that counterbalanced technical problem solving in the field. After Americans landed on the Moon, new executives established a more decentralized system. By moving program control out of Washington, they hoped to strengthen NASA's technical capability. Field officials fought for the new system because it gave them more control over their own future.

In this theory on the life cycle of bureaus, Anthony Downs suggests that organizational aging will occur as conservers gain the upper hand in the struggle for power with agency innovators.[26] In NASA, struggles for power occurred. The struggles, however, took place between different factions within the overall culture rather than different personalities. The resolution of the management issue had more to do with cultural politics than aging per se.

As a confederation of cultures, NASA contained people whose organizational philosophies often differed. Controversies arose during the first decade over issues such as all-up versus incremental testing, the proper role of contractors, and the amount of central control needed to direct field center work on expeditions like Project Apollo. Organizational politics helped to resolve such controversies. Different factions advanced their positions, which were usually resolved on the basis of individual skill and organizational goals. Although dominant positions emerged, losing factions still held to their points of view. The losing side could wait until changing conditions permitted reconsideration of the dominant mode. NASA did not function as a cultural melting pot so much as a political stew.[27]

Given the imperatives of the Gemini and Apollo space flight programs, the subculture of strong central control prevailed during the 1960s. As the Apollo program wound down, and NASA began the development of the space shuttle, the balance of power shifted. The Air Force group that promoted the doctrine of central control during the first generation was not as well situated to shape the management plan for the space shuttle, nor was the mandate for the shuttle as demanding as that for the race to the Moon. Factions from the field centers responsible for the development of the space shuttle prevailed in their effort to establish a much looser system of supervision. The looser system of supervision did not work as well. Rather than simplify the management of new programs like the space shuttle or the space station *Freedom,* the new systems complicated it.[28] Organizational aging played a role in this transformation to the extent that factions could no longer rely solely upon the importance of the mission to resolve such disputes. Field officials had to worry about their survival and technical capability as well. During the Apollo era, NASA officials resolved problems like this by addressing the question, "Do we want to get to the Moon or not?"[29] That problem-solving orientation waned as NASA aged, and the imperatives of Project Apollo gave way to other concerns.

Life cycle theorists observe that government agencies become more concerned with survival as they mature and they predict that agencies will transform their methods of doing business as a result. NASA officials had to balance the importance of missions against the needs of embattled field centers to gain a fair share of new programs and protect themselves against outside control. The management of NASA changed as a result.

Culture and Performance

During its first decade, NASA exhibited many of the characteristics scholars associate with organizations that perform very well. It balanced technical discretion with central control. It developed a tradition of open communication and flexibility when faced with new challenges. It had a clear sense of mission and a strong organizational culture.

In government, it appears, characteristics such as these are hard to sustain. They do not persist for extended periods of time. The NASA experience suggests that three forces work against the maintenance of a high-performance culture in government: a volatile political environment, the long-term trend toward an increasing administrative burden in the government as a whole, and the natural processes of aging, fed by the inevitable expansion and contraction cycle that government bureaus go through. None of these turn out to be forces over which the managers of affected agencies can exerise much control.

These forces push high-performance government organizations toward the bureaucratic mold. Now, a bureaucracy can be a wonderful form of organization for carrying out some activities. Bureaucracies mail out checks on time, deliver supplies to troops in the field, and regulate commercial affairs. They do this by employing career civil servants who institute written rules through an official hierarchy that reduces the opportunity for favoritism or arbitrary decisions.[30]

The most successful science and technology programs of the twentieth century, however, have managed in some fashion to create a culture that circumvents the bureaucratic mode. Government leaders do this, in general, by isolating the agency as much as possible from the forces that weaken the high-performance mode. They create crash programs that exist for only short periods of time. The Manhattan Project that developed the atom bomb and the National Radiation Laboratory project for the development of radar during World War II disappeared before bureaucratic imperatives could take hold. They create programs with a strong sense of mission, one of the principal ways for motivating employees to work hard and fight off forces that compromise organizational effectiveness.[31] During the race to the Moon, NASA possessed such a mission.

Designers of high-performance organizations place them well outside the bureaucratic mainstream, removed from official procedures and constrictions. Naval officers located the Polaris missile development program in a special projects office under the Secretary of the Navy, separated from the regular Navy bureaus.[32] The radar development project during World War II was run through the National Radiation Laboratory, a special organization set up at the Massachusetts Institute of Technology. The Synthesis Group, established in 1990 to suggest a strategy for returning to the Moon and moving on to Mars, concluded that a new National Program Office would be necessary to accomplish it. "The Space Exploration Initiative is so great in scope," the group ventured, "that it cannot be executed in a 'business as usual' manner and have any chance for success."[33]

None of these strategies provide such organizations with the capability to enter the governmental mainstream and sustain high-performance cultures for extended periods of time. Crash programs, specific missions, and special organizations are all by their nature methods for avoiding governmental routine. The NASA experience suggests that any attempt to retain a high-performance culture inside the governmental mainstream is doomed to fail.

The beliefs and assumptions underlying high-performance cultures may persist for some time, even after their original mission fades. As puzzling as this may seem, NASA engineers and scientists of the second generation retained many beliefs from the first generation even as they lost the ability to practice them. This may testify to the natural tendency of NASA engineers and scientists to approach their work as professionals rather than as bureaucrats. It may reflect an ability among first-generation officials to recruit second-generation professionals that fit the agency mold or to "break them in" in some unofficial way. By whatever method, NASA officials passed on their first-generation beliefs to the second generation. They were unable, however, to pass on their ability to translate those beliefs into practice. A culture that permits such a gap between belief and practice cannot persist for long.

The early NASA culture developed out of institutions that had been practicing their beliefs for some period of time. The joining of values and practices created powerful traditions on which early NASA officials drew. "Look," said one of the first-generation engineers who had watched those traditions mature for more than a dozen years in a predecessor organization, "I don't know what makes this damn thing go that we've got here—that makes us so good and so great, and our capability to inspire people. But I'll tell you this, if we ever lose it I won't know how to tell you to regain it."[34] However short-lived they may be, successful cultures are not created overnight. NASA's early culture developed for many years before it blossomed during the first decade of space exploration. The period of transformation that followed took place over two decades. If NASA is to fashion a culture that is both successful and mature, it too will be years in developing.

Appendix:
NASA Culture Survey

A survey was administered to a random sample of eight hundred NASA professional employees (engineers, scientists, and professional administrators) in the summer of 1988. Seven hundred and four employees returned the survey. For questions used to measure change in beliefs, scores from the first generation of NASA employees (those who joined NASA prior to 1970) are compared to those from the second generation (those who joined in 1970 and later). Responses are in percentages; they do not always add up to 100 percent because of rounding. The survey question along with its number is given in the tables that follow.

Recruiting Exceptional People

	Year Joined NASA[a]		
Opinion	1951–1969	1970–1988	All
2. NASA still recruits exceptional people.[b]			
Strongly agree	5	4	5
Agree	44	43	44
No opinion; undecided	24	23	24
Disagree	23	28	25
Strongly disagree	4	2	3
1. When I first joined the agency, NASA recruited truly exceptional people.[c]			
Strongly agree	30	8	19
Agree	51	54	52
No opinion; undecided	10	20	15
Disagree	9	17	13
Strongly disagree	1	1	1

[a]NASA was created in 1958. Respondents who worked for the organizations out of which NASA was formed marked the date they joined them.
[b]Not statistically significant at the .01 level.
[c]Statistically significant at the .01 level.

Working for Industry

	Year Joined NASA		
Response	1951–1969	1970–1988	All
40. Prior to joining NASA, did you work for more than two years as a full-time, permanent employee in industry?[a]			
Yes	48	36	42
No	52	65	58

[a]Statistically significant at the .01 level.

Attitudes toward Risk and Failure

Opinion	All
18. Risk and failure are a normal part of the business of developing new technologies.[a]	
Strongly agree	39
Agree	58
No opinion; undecided	1
Disagree	2
Strongly disagree	<1
19. NASA employees are allowed to fail and learn from their mistakes.	
Strongly agree	2
Agree	44
No opinion; undecided	23
Disagree	26
Strongly disagree	5
23. NASA rewards people who are technically creative.	
Strongly agree	6
Agree	45
No opinion; undecided	19
Disagree	24
Strongly disagree	7
20. NASA has stayed on the cutting edge of new technologies.	
Strongly agree	6
Agree	42
No opinion; undecided	17
Disagree	30
Strongly disagree	5

Opinion	All

22. Cost constraints have forced us to cut corners in carrying out our programs.

Strongly agree	33
Agree	47
No opinion; undecided	10
Disagree	9
Strongly disagree	1

[a]No statistically significant difference between first- and second-generation employees at the .01 level.

Attitudes toward Management

Opinion	All

26. The time we spend on management tends to be time taken away from basic engineering and science.

Strongly agree	20
Agree	47
No opinion; undecided	13
Disagree	18
Strongly disagree	1

27. Engineers make good managers.

Strongly agree	4
Agree	28
No opinion; undecided	34
Disagree	26
Strongly disagree	9

28. Scientists make good managers.

Strongly agree	1
Agree	14
No opinion; undecided	38
Disagree	35
Strongly disagree	12

Attitudes toward Change

Opinion	Year Joined NASA		All
	1951–1969	1970–1988	

8. NASA should concentrate on implementing programs already approved rather than pushing for new programs.[a]

Strongly agree			4
Agree			16
No opinion; undecided			18
Disagree			47
Strongly disagree			16

11. I am very optimistic about NASA's future.[a]

Strongly agree			11
Agree			42
No opinion; undecided			23
Disagree			20
Strongly disagree			4

9. Most of the changes that I have seen take place within NASA have made the agency work better.[b]

Strongly agree	1	1	1
Agree	15	19	17
No opinion; undecided	20	32	26
Disagree	50	41	46
Strongly disagree	15	6	10

10. People within NASA welcome change.[b]

Strongly agree	6	1	3
Agree	38	31	34
No opinion; undecided	24	24	24
Disagree	30	37	34
Strongly disagree	2	6	4

[a]No statistically significant difference between first- and second-generation employees at the .01 level.
[b]Statistically significant at the .01 level.

	Year Joined NASA		
Response	1951–1969	1970–1988	All

37. While in high school, in what sort of community did your family live?[a]

An affluent residential neighborhood in or near a city	5	11	8
A residential neighborhood in or near a city	31	44	38
A city neighborhood of working people	21	12	17
In a small town	27	24	25
In the country	15	10	12

38. Did your parents contribute more than half of the cost of your undergraduate college education?[b]

Yes	43	47	45
No	57	54	55

[a]Statistically significant at the .01 level.
[b]Not statistically significant at the .01 level.

Perceptions of the NASA Bureaucracy

Opinion	All

12. NASA places a great deal of emphasis on paperwork and procedures.

Strongly agree	28
Agree	56
No opinion; undecided	7
Disagree	9
Strongly disagree	<1

14. It is relatively easy to cut through the bureaucracy and get things done within NASA today.

Strongly agree	<1
Agree	11
No opinion; undecided	12
Disagree	56
Strongly disagree	21

Opinion	All

16. It is fairly easy to change official procedures within NASA once they are approved.

Strongly agree	<1
Agree	9
No opinion; undecided	24
Disagree	50
Strongly disagree	17

The NASA Bureaucracy over Time

	Year Joined NASA		
Opinion	1951–1969	1970–1988	All

13. The amount of paperwork has increased substantially since I came to work for NASA.[a]

Strongly agree	45	22	33
Agree	36	35	36
No opinion; undecided	4	25	15
Disagree	14	18	16
Strongly disagree	1	1	1

15. It was much easier to cut through bureaucratic barriers and get things done when I first came to the agency.[a]

Strongly agree	30	3	16
Agree	45	27	36
No opinion; undecided	10	40	25
Disagree	13	26	20
Strongly disagree	3	3	3

[a]Statistically significant at the .01 level.

Managerial Risk Taking and Communication

	Year Joined NASA		
Opinion	1951–1969	1970–1988	All

24. At the management level, NASA is dominated by people who are cautious and inclined to avoid risks.

Strongly agree	16
Agree	46
No opinion; undecided	22
Disagree	16
Strongly disagree	1

| | Year Joined NASA | | |
Opinion	1951–1969	1970–1988	All

25. At the management level, the number of people who are cautious and inclined to avoid risks has increased since I joined the agency.[a]

Strongly agree	21	10	16
Agree	45	32	39
No opinion; undecided	18	44	31
Disagree	15	13	14
Strongly disagree	1	<1	1

17. People within the agency—including those at different centers—communicate with each other as much as they should.

Strongly agree			1
Agree			17
No opinion; undecided			18
Disagree			47
Strongly disagree			18

[a]Statistically significant at the .01 level.

In-House Capability

| | Year Joined NASA | | |
Opinion	1951–1969	1970–1988	All

5. NASA has turned over too much of its basic engineering and science work to contractors.[a]

Strongly agree			31
Agree			47
No opinion; undecided			11
Disagree			11
Strongly disagree			1

4. Since I came to work for the agency, NASA has lost much of its in-house technical capability. (Respondents were told to consider only the technical capability of NASA employees—not the contributions of support contractors.)[b]

Strongly agree	30	11	20
Agree	48	39	43
No opinion; undecided	5	23	14
Disagree	13	24	18
Strongly disagree	4	4	4

[a]No statistically significant difference between first- and second-generation employees at the .01 level.
[b]Statistically significant at the .01 level.

NASA Testing Activities

Opinion	Year Joined NASA		
	1951–1969	1970–1988	All
6. NASA did substantially more testing in the past than it does today.[a]			
Strongly agree	26	10	18
Agree	40	29	34
No opinion; undecided	19	49	34
Disagree	14	11	12
Strongly disagree	2	2	2
7. NASA should do more testing than it currently does.[b]			
Strongly agree			18
Agree			41
No opinion; undecided			30
Disagree			10
Strongly disagree			1

[a]Statistically significant at the .01 level.
[b]No statistically significant difference between first- and second-generation employees at the .01 level.

Interactive Failures

Opinion	All
21. As the technologies we use have become more complex, the possibility of an interactive failure on a manned or unmanned space flight has increased. (Respondents were told that an interactive failure was "one in which two or more single point failures combine to create a more serious problem.")	
Strongly agree	12
Agree	49
No opinion; undecided	23
Disagree	14
Strongly disagree	1

Educational Backgrounds of NASA Employees

	Year Joined NASA		
Response	1951–1969	1970–1988	All

34. When you earned your bachelor's degree, where do you think you placed among your classmates in the same major?[a]

Top 5%	12	24	18
Top 10%	28	29	29
Top 25%	33	31	32
Top 50%	20	14	17
Bottom 50%	7	2	4

35. Below is a list of colleges and universities.[b] *Did you earn a degree (graduate or undergraduate) from any of the schools named on the list?*[c]

Yes	9	9	9
No	91	91	91

[a]Statistically significant at the .01 level.
[b]California Institute of Technology, Cornell University, Princeton University, Purdue University at Lafayette, Massachusetts Institute of Technology, Stanford University, University of California at Berkeley, University of Illinois at Urbana, University of Michigan at Ann Arbor.
[c]Not statistically significant at the .01 level.

Work of the NASA Professional Employee

Response	All

33. Would you say that you spend most of your work week:

Working in a laboratory, test facility, control or tracking center; training astronauts; or working on space flight or aeronautics hardware (including satellites and space probes)	3
Working at a desk in an office	77
Doing both	20

32. We need to know how much time the average technical or professional NASA employee spends supervising contractors. Would you say that you presently spend:

A great deal of time supervising contractors	31
Some, but not a great deal of time	46
None or practically none at all	23

Hands-On Work

Opinion	Year Joined NASA		
	1951–1969	1970–1988	All

29. As NASA employees, we have as much opportunity to do "hands-on" work as we want.[a]

Strongly agree	1	4	2
Agree	19	22	21
No opinion; undecided	12	13	13
Disagree	58	47	52
Strongly disagree	10	14	12

30. When I first came to the agency, we did a lot more "hands-on" work than we do today.[a]

Strongly agree	28	6	17
Agree	56	32	44
No opinion; undecided	8	42	25
Disagree	8	19	13
Strongly disagree	0	1	1

[a]Statistically significant at the .01 level.

The Lure of Space Exploration

Opinion	All

31. Working on the space program is still just as exciting as I thought it would be.

Strongly agree	15
Agree	47
No opinion; undecided	12
Disagree	23
Strongly disagree	3

Faith in Technical Capability

Opinion	All

3. NASA has just as much technical capability as its contractors.[a]

Strongly agree	19
Agree	40
No opinion; undecided	11
Disagree	26
Strongly disagree	4

[a]No statistically significant difference between first- and second-generation employees at the .01 level.

Essay on Sources

During NASA's formative years, people associated with the American space program did not speak much about their organizational culture. They talked about the challenges of space exploration. They described the technology required to conduct expeditions. They outlined the progress of specific missions. Occasionally they wrote about the management techniques they adopted. *Culture*, however, was not a word frequently employed. The people who ran the American space program were not inclined at that time to reflect extensively upon the organizational philosophy that made their accomplishments possible.

As a result, primary sources from the early years of space flight describing what people would later recognize as NASA's organizational culture rarely appeared. A few materials, however, do exist. A series of articles dealing in part with the organizational philosophy of the manned space flight program appeared in the March 1970 issue of *Astronautics & Aeronautics*. Eight of the articles were reissued as a NASA special publication under the title *What Made Apollo a Success?* SP-287 (Washington: NASA, 1971). A NASA special publication edited by Edgar M. Cortright, *Apollo Expeditions to the Moon*, SP-350 (Washington: NASA, 1975), deals primarily with engineering decisions and space flight operations. Since the chapters were written by NASA managers who actually carried out the missions, this work contains insights into the thoughts of those people at that time.

The most extensive testimony by NASA space flight managers and their industrial contractors that touches on early organizational philosophy can be found in congressional hearings. Statements on the original philosophy of the NASA space program are contained in hearings held by the House Committee on Science and Astronautics, *Review of the Space Program*, 86th Cong., 2d sess., pt. 1, 1960. In 1964 the House Committee on Science and Astronautics, Subcommittee on Manned Space Flight, published briefings conducted at NASA space flight centers and contractor facilities in the *1965 NASA Authorization*, 88th Cong., 2d sess., pt. 2, 1964. House Committee on Science and Astronautics, *1968 NASA Authorization*, 90th Cong., 1st sess., pt. 1, 1967, contains hearings on the accident at Launch Complex 34 that took three astronauts' lives and includes reflections on NASA's approach to risk taking. Failures that led to the *Investigation of Project Ranger*, 88th Cong., 2d sess., 1964, conducted by the House Committee on Science and Astronautics, Subcommittee on NASA Oversight, produced some insights into the operational philosophy of the unmanned program. In 1969, staff members of the House Committee on Science and Astronautics, Subcommittee on NASA Oversight, published comments by contractors and NASA managers in *Apollo Program Management*, 91st Cong., 1st sess.

Secondary sources are more extensive. NASA conducts an ambitious history program (of which this study is a part). The NASA History Office has sponsored histories of every field center and nearly every large U.S. civilian space undertaking. Among the histories that were especially useful in tracing NASA's early organizational culture were the study of the Langley Aeronautical Laboratory before it became part of NASA, by James R. Hansen, *Engineer in Charge*, SP-4305 (Washington: NASA, 1987), and the history of the German rocket team working to produce the Apollo/Saturn launch vehicle, written by Roger E. Bilstein, *Stages to Saturn*, SP-4206 (Washington: NASA, 1980). Virginia P. Dawson has written an insightful history of the Lewis Research Center, *Engines and Innovation: Lewis Laboratory and American Propulsion Technology*, SP-4306 (Washington: NASA, 1991), and Richard P. Hallion has traced the development of the high-speed flight research station, *On the Frontier: Flight Research at Dryden, 1946–1981*, SP-4303 (Washington: NASA, 1984). The early development of the Kennedy Space Center is described by Charles D. Benson and William B. Faherty in *Moonport: A History of Apollo Launch Facilities and Operations*, SP-4204 (Washington: NASA, 1978), and Henry C. Dethloff has prepared *Suddenly Tomorrow Came: A History of the Johnson Space Center, 1962–1990*, SP-4307 (Washington: NASA, 1992). A history of the Marshall Space Flight Center by Andrew J. Dunar was under way while this book was being written.

Three books provide a comprehensive history of NASA's early human space flight program: Loyd S. Swenson, James M. Grimwood, and Charles C. Alexander, *This New Ocean: A History of Project Mercury*, SP-4201 (Washington: NASA, 1966); Barton C. Hacker and James M. Grimwood, *On the Shoulders of Titans: A History of Project Gemini*, SP-4203 (Washington: NASA, 1977); and Courtney G. Brooks, James M. Grimwood, and Loyd S. Swenson, *Chariots for Apollo: A History of Manned Lunar Spacecraft*, SP-4205 (Washington: NASA, 1979). Arnold S. Levine's book, *Managing NASA in the Apollo Era*, SP-4102 (Washington: NASA, 1982), and Robert L. Rosholt's earlier study, *An Administrative History of NASA, 1958–1963*, SP-4101 (Washington: NASA, 1966), although not presented as culture studies, contain useful material on administrative practices. Robert W. Smith illuminates NASA's transformational problems in *The Space Telescope: A Study of NASA Science, Technology, and Politics* (New York: Cambridge University Press, 1989). A history by Sylvia D. Fries, drawing on the recollections of first-generation NASA engineers, contains much material on NASA's original culture, *NASA Engineers and the Age of Apollo*, SP-4104 (Washington: NASA, 1992). Available in manuscript form, it provided much of the inspiration for this study of cultural change.

Outside histories also proved useful. Two of the best are Charles Murray and Catherine Bly Cox, *Apollo: The Race to the Moon* (New York: Simon and Schuster, 1989), which describes the culture of the engineers who designed and conducted the flights to the Moon; and Tom Wolfe, *The Right Stuff* (New York: Farrar, Straus, Giroux, 1979), on the culture of the astronauts. Frederick I. Ordway and Mitchell R. Sharpe's *The Rocket Team* (New York: Thomas Y. Crowell, 1979) traces the origins of the group that formed the nucleus of NASA's Marshall Space Flight Center.

Special reports often contain insights into institutional philosophy, especially those generated in response to a crisis. The report of the Rogers Commission (Presidential Commission on the Space Shuttle Challenger Accident), *Report of the Presidential Commission* (Washington: Government Printing Office, 1986), is a rich source of material on the operational norms of the NASA space flight program in the mid-1980s. The report of the Augustine Committee (the Advisory Committee on the Future of the U.S. Space Program), *Report of the Advisory Committee* (Washington: Government Printing Office, 1990), deals explicitly with NASA's organizational culture. A report produced by the National Academy of Public Administration, "Maintaining the Program Balance," vol. 1 (1991), deals with issues of in-house capability and contractor responsibility raised in the course of NASA's cultural journey.

The work on organizational culture in general is voluminous. Readers just starting to become acquainted with this perspective would do well to begin with the survey by J. Stephen Ott, *The Organizational Culture Perspective* (Chicago: Dorsey, 1989), or the overview by Edgar H. Schein, *Organizational Culture and Leadership* (San Francisco: Jossey-Bass, 1985). Ongoing issues are analyzed in the anthology by Peter J. Frost and Larry F. Moore, *Reframing Organizational Culture* (Newbury Park, Calif.: Sage, 1991). The organizational culture literature spawned (and was spawned by) frequent attempts to identify cultural norms exhibited by high-performing organizations. Three seminal works in the organizational excellence field are W. G. Ouchi, *Theory Z: How American Business Can Meet the Japanese Challenge* (Reading, Mass.: Addison-Wesley, 1981); Thomas J. Peters and Robert H. Waterman, *In Search of Excellence* (New York: Harper & Row, 1982); and Rosabeth Moss Kanter, *The Change Masters* (New York: Simon & Schuster, 1983).

The organizational culture literature draws mainly on the experience of business firms, although books have been written that describe the cultures of government bureaus. One of the best to analyze the culture of a government agency (without calling it a culture) is the classic study by Herbert Kaufman, *The Forest Ranger* (Baltimore: Johns Hopkins Press, 1960). A project to study high-reliability organizations addresses cultural issues arising in the public sector. See Todd R. LaPorte and Paula M. Consolini, "Working in Practice but Not in Theory: Theoretical Challenges of 'High-Reliability Organizations,'" *Journal of Public Administration Research and Theory* 1 (January 1991): 19–47; and Karl E. Weick, "Organizational Culture as a Source of High Reliability," *California Management Review* 24 (Winter 1987): 112–27. Insightful summaries of public agency cultures can be found in chapter 8 of Harold Seidman and Robert Gilmour, *Politics, Position, and Power,* 4th ed. (New York: Oxford University Press, 1986), and in chapter 6 of James Q. Wilson, *Bureaucracy: What Government Agencies Do and Why They Do It* (New York: Basic, 1989).

A moderate amount of scholarship has been produced on the ways that organizations, especially those in the government, change over time. John R. Kimberly, Robert H. Miles, and associates summarize much of this work in *The Organizational Life Cycle* (San Francisco: Jossey-Bass, 1987), including three chapters on organizational decline. The notion that organizations move through life cycles has fascinated scholars for some time. Anthony Downs offers a theory to explain the "life cycle of bureaus" in chapter 2 of *Inside Bureaucracy* (Boston: Little, Brown, 1966), while Marver Bernstein sets forth a life-cycle theory of regulatory agencies in *Regulating Business by Independent Commission* (Princeton: Princeton University Press, 1955). Kenneth J. Meier and John P. Plumlee tested the life-cycle thesis in "Regulatory Administration and Organizational Rigidity," *Western Political Quarterly* 31 (March 1978): 80–95.

Research for this study of NASA's changing organizational culture began in the summer of 1987. In contrast to the paucity of insights from people speaking for the record during the 1960s, NASA managers in the 1980s spoke earnestly of the agency's organizational culture. In many cases, these were the same people who had managed NASA programs in the earlier era. Perhaps their ability to philosophize was due to a growing awareness of the importance of organizational culture in management circles generally. Perhaps it was due to their sense that NASA's culture had changed, especially after the trauma engendered by the explosion of the space shuttle *Challenger* in early 1986. Either way, NASA managers who had been tight-lipped about their organizational philosophy in the 1960s seemed anxious to talk about it in the 1980s. Interviews with NASA managers provided much of the material for this study.

Subjects were selected for interview through a position identification method. Put simply, the top career service positions at NASA Headquarters and the field centers were identified. The people who held those positions during the mid-1960s were located and interviewed, as was a counterpart group holding equivalent positions during the mid-1980s (before the *Challenger* accident). Most of the former had retired from NASA.

During the interviews, all subjects were asked to describe their personal backgrounds and their work with the civilian space program. They were asked to describe NASA generally and to discuss any ways in which it had changed. They were asked questions about the use of technology, the management of risk, the funding of agency programs, and the people with whom they worked. A five-page list of questions guided the interview, although each interview was conducted in the form of a general discussion. Subjects were told that their names would not be placed next to any quotations used in this book but that a transcript of their interview would make its way to the NASA History Office. This was done to provide a modicum of anonymity for those persons who still worked for NASA at the time of their interview. All of the interviews were conducted in person by the author. Each interview took approximately ninety minutes to complete.

A list of the positions of the persons interviewed for this study follows. When quoted in the text, these interviews are cited in the notes with a number. Additional interviews, undertaken by the author and other researchers for other studies, are cited by name and date. All of the interviews cited are available through the NASA History Office in Washington, D.C. Researchers may locate numbered interviews by consulting a key at the NASA History Office.

Associate Administrator, Manned Space Flight, Headquarters, 1965
Director, Apollo Program, Headquarters, 1965
Director, Physics and Astronomy Programs, Headquarters, 1965
Director, Lunar and Planetary Programs, Headquarters, 1965
Assistant Administrator for Administration, Headquarters, 1965
Associate Administrator, Office of Space Flight, Headquarters, 1985
Associate Administrator, Office of Space Station, Headquarters, 1985
Associate Administrator for Management, Headquarters, 1985
Director, Manned Spacecraft Center, 1965
Director of Flight Operations, MSC, 1965
Director of Engineering and Development, MSC, 1965
Manager, Apollo Spacecraft Program, MSC, 1965
Director, Johnson Space Center, 1985
Director, Mission Operations, JSC, 1985
Director, Flight Crew Operations, JSC, 1985
Deputy Director (Technical), Marshall Space Flight Center, 1965
Director, Astrionics Laboratory, MSFC, 1965
Director, Test Division, MSFC, 1965
Manager, Saturn I/IB Program Office, MSFC, 1965
Director, Marshall Space Flight Center, 1985

Director, Shuttle Projects Office, MSFC, 1985
Director, Launch Operations, Kennedy Space Center, 1966
Director, Design Engineering, KSC, 1967
Director, Kennedy Space Center, 1985
Director, Launch and Landing Operations, KSC, 1985
Director, Shuttle Engineering, KSC, 1985
Director, Langley Research Center, 1968
Assistant Director, Aeronautical Research, LaRC, 1965
Assistant Director, Group 1, LaRC, 1968
Assistant Director, Group 2, LaRC, 1965
Director, Langley Research Center, 1984
Director for Electronics, LaRC, 1985
Director for Structures, LaRC, 1985
Assistant Director, Aeronautics and Flight Mechanics,
    Ames Research Center, 1965
Director of Astronautics, ARC, 1966
Director, Lewis Research Center, 1965
Director, Lewis Research Center, 1985
Director, Goddard Space Flight Center, 1965
Director of Space and Earth Sciences, GSFC, 1965
Project Manager, Orbiting Astronomical Observatory, GSFC, 1966
Director, Goddard Space Flight Center, 1985
Director, Space and Earth Sciences, GSFC, 1985
Director, Flight Research Center, 1965
Chief, Flight Operations, FRC, 1966
Chief, Research Aircraft Operations, Dryden Flight Research Facility, 1985

The persons interviewed for this study drew a fairly consistent picture of NASA's operational philosophy and the ways it had changed. Additional sources were consulted to address two questions. First, were the views expressed by top-level officials widely held by professional employees in the agency? One might assume that top-level people embodied NASA's dominant norms, since they had advanced up the institutional hierarchy. To check this assumption, a survey was administered by the author to a random sample of 800 NASA engineers, scientists, and professional administrators. The sample was drawn from a computerized list of the 15,660 employees classified by NASA personnel managers as "professional employees" as of August 1988. Clerical, wage and craft, and technical support employees were not included in the survey. Professional employees make up over 70 percent of all permanent employees on the NASA payroll, who in 1988 numbered nearly 22,000. Seven hundred and four persons returned the survey, for a response rate of 88 percent. Survey results are presented in the appendix. The survey results closely match the impressions gathered during the interviews, suggesting a good degree of consensus about NASA's overall culture and the changes that have taken place in it.

The second question concerning the interview results proved more difficult to address. Subjects were asked in the late 1980s to identify their 1960s organizational culture. Were the cultural norms they identified actually in effect during the 1960s, when the subjects worked for NASA? Nostalgia and the passage of time can cloud recollections. Worse still, beliefs about reality may be subject to misinterpretation at any time. How was it possible to tell whether a recollected cultural norm, widely professed by NASA employees, was a historical reality? Cultural norms, of course, do not have to be objectively true in order to motivate behavior. Nonetheless, sorting objective realities from those imagined or held solely on faith proved to be an important challenge in this study.

A number of sources were used to cross-check beliefs and historical reality. Formal records and statistics were examined. The Human Resources and Education Office at NASA headquarters annually prepares a report, "The Civil Service Work Force." The NASA History Office has produced a three-volume set of factual data on NASA activities. Jane Van Nimmen and Leonard C. Bruno with Robert L. Roshold, *NASA Historical Data Book*, SP-4012 (Washington: NASA, 1988), provide an avalanche of statistics on NASA activities from 1958 to 1968. Linda Neuman Ezell, *NASA Historical Data Book*, vols. 2 and 3, SP-4012 (Washington: NASA, 1988), presents data on programs and projects from 1958 to 1978. The tiny *Pocket Statistics*, published annually by NASA, provides summary data on a wide range of subjects. *Origins of NASA Names*, by Helen T. Wells, Susan H. Whiteley, and Carrier E. Karegeannes, SP-4402 (Washington: NASA, 1976), summarizes NASA programs and provides short histories of each NASA installation.

Primary and secondary sources were consulted to determine whether the observations made by persons looking back at the 1960s with twenty years of perspective differed from those of NASA officials at that time. Testimony by NASA officials at congressional hearings during the 1960s proved especially useful, as did the many books prepared through the auspices of the NASA History Office. Interviews conducted for other NASA history projects were examined. Since in many cases the same NASA officials were speaking, it was possible to check perceptions across time. Nearly all of these materials can be found in the archives maintained by the NASA History Office at NASA Headquarters in Washington, D.C.

Finally, a small group of outside experts was interviewed. These were people familiar with NASA operations (their names were drawn from the membership lists of NASA advisory committees), but none had ever worked as part of the NASA civil service. They were asked to address the same topics that the NASA insiders commented upon in their interviews. These outside experts were promised anonymity (most worked for institutions that received NASA contracts), and no transcripts of their remarks were prepared. A list of the institutions for which these individuals worked follows.

The Boeing Company
The Lockheed Corporation
TRW Inc.
McDonnell Douglas Corporation
Systems Technology, Inc.
Harvard-Smithsonian Center for Astrophysics
University of California at Los Angeles
AT&T Bell Laboratories

Ultimately, I had to rely on the words of NASA officials to fathom their organizational culture. In many ways, this is their history. Some scholars believe that cultural norms are so deeply held that employees cannot articulate them. In this study, that did not turn out to be the case. Not only did most people interviewed between 1987 and 1991 understand the meaning of culture, they spoke easily about the culture of the civilian space agency and the ways in which they had seen it change.

Notes

Introduction: NASA's Organizational Culture

Epigraph: Richard Corrigan, "NASA's Midlife Crisis," *National Journal* 18 (March 22, 1986): 686.

1. See Marver H. Bernstein, *Regulating Business by Independent Commission* (Princeton: Princeton University Press, 1955), and Anthony Downs, *Inside Bureaucracy* (Boston: Little, Brown, 1967). For a contrary perspective, see Herbert Kaufman, *Time, Chance, and Organization: Natural Selection in a Perilous Environment*, 2d ed. (Chatham, N.J.: Chatham House, 1991).

2. Walter A. McDougall, . . . *The Heavens and the Earth: A Political History of the Space Age* (New York: Basic, 1985).

3. Robert W. Smith, *The Space Telescope: A Study of NASA Science, Technology, and Politics* (New York: Cambridge University Press, 1989).

4. Presidential Commission on the Space Shuttle Challenger Accident (William P. Rogers, chairman), *Report of the Presidential Commission* (Washington: Government Printing Office, 1986).

5. "Space Shuttle Program: Statement by the President Announcing the Decision to Proceed with Development of the New Space Transportation System," *Weekly Compilation of Presidential Documents* (January 5, 1972): 27–28; Office of the White House Press Secretary, "Press Conference of Dr. James Fletcher and George M. Low, San Clemente Inn, California," January 5, 1972, NASA History Office, Washington.

6. Council for Excellence in Government, "Federal Reserve, NIH and NSC rated as Most Respected Government Agencies," November 6, 1990, Council for Excellence in Government, Washington.

7. J. Steven Ott, *The Organizational Culture Perspective* (Chicago: Dorsey, 1989), 50; also see 70–73.

8. See Philip Selznick, *TVA and the Grass Roots* (New York: Harper & Row, 1949); Morton H. Halperin, *Bureaucratic Politics and Foreign Policy* (Washington: Brookings, 1974); and James Q. Wilson, *Bureaucracy: What Government Agencies Do and Why They Do It* (New York: Basic, 1989).

9. Herbert Kaufman, *The Forest Ranger* (Baltimore: Johns Hopkins Press, 1960), 197, 200.

10. William G. Ouchi, *Theory Z: How American Business Can Meet the Japanese Challenge* (Reading, Mass.: Addison-Wesley, 1981).

11. Thomas J. Peters and Robert H. Waterman, *In Search of Excellence* (New York: Harper & Row, 1982).

12. W. Brooke Tunstall, "The Breakup of the Bell System: A Case Study in Cultural Transformation," in *Gaining Control of the Corporate Culture*, edited by Ralph H. Kilmann, Mary J. Saxton, and Roy Serpa (San Francisco: Jossey-Bass, 1985); Lee Iacocca with William Novak, *Iacocca: An Autobiography* (Toronto: Bantam, 1984).

13. Karl E. Weick, "Organizational Culture as a Source of High Reliability," *California Management Review* 24 (Winter 1987): 112–27; Todd R. LaPorte and Paula M. Consolini, "Working in Practice but Not in Theory: Theoretical Challenges of 'High-Reliability Organizations,'" *Journal of Public Administration Research and Teaching* 1 (January 1991): 19–47.

14. Daniel T. Carroll, "A Disappointing Search for Excellence," *Harvard Business Review* (November/December 1983): 78–88.

15. Ott, *The Organizational Culture Perspective.*

16. Ibid. Chap. 9.

17. Edgar H. Schein describes this method in "How to Uncover Cultural Assumptions in an Organization," in Schein, *Organizational Culture and Leadership* (San Francisco: Jossey-Bass 1985).

Chapter 1: Building Blocks

Epigraph: Harold Seidman and Robert Gilmour, *Politics, Position, and Power* (New York: Oxford University Press, 1970), 98–99.

1. Public Law 271, 63d Cong., 3d sess., passed March 3, 1915 (38 Stat. 930). For this and other legislation, see Alex Roland, *Model Research: The National Advisory Committee for Aeronautics, 1915–1958*, SP-4103, vol. 2 (Washington: NASA, 1985), 393–422.

2. James R. Hansen, *Engineer in Charge: A History of the Langley Aeronautical Laboratory, 1917–1958*, SP-4305 (Washington: NASA, 1987).

3. Elizabeth A. Meunger, *Searching the Horizon: A History of Ames Research Center, 1940–1976*, SP-4304 (Washington: NASA, 1985).

4. Virginia P. Dawson, *Engines and Innovation: Lewis Laboratory and American Propulsion Technology*, SP-4306 (Washington: NASA, 1991), 8–9, 23; Roland, *Model Research*, vol. 1, 88.

5. Richard P. Hallion, *On the Frontier: Flight Research at Dryden, 1946–1981*, SP-4303 (Washington: NASA, 1984), 360.

6. Hugh L. Dryden, "Space Technology and the NACA," speech prepared for the luncheon meeting of the Institute of the Aeronautical Sciences, New York, January 27, 1958, 7, NASA History Office, Washington.

7. Jacob Neufeld, *The Development of Ballistic Missiles in the United States Air Force, 1945–1960* (Washington: USAF Office of Air Force History, 1990), 125–28, 140, 210–12.

8. Constance M. Green and Milton Lomask, *Vanguard: A History* (Washington: Smithsonian Institution Press, 1971).

9. Frederick I. Ordway and Mitchell R. Sharpe, *The Rocket Team* (New York: Thomas Y. Crowell, 1979).

10. Wernher von Braun, "Crossing the Last Frontier," *Collier's* 129 (March 22, 1952).

11. Presidential Commission on the Space Shuttle Challenger Accident, *Report of the Presidential Commission*, 104.

12. Jane Van Nimmen and Leonard C. Bruno, *NASA Historical Data Book,* vol. 1, *NASA Resources 1958–1968,* SP-4012 (Washington: NASA,1988), 332.

13. Charles D. Benson and William B. Faherty, *Moonport: A History of Apollo Launch Facilities and Operations,* SP-4204 (Washington: NASA, 1978).

14. Loyd S. Swenson, James M. Grimwood, and Charles C. Alexander, *This New Ocean: A History of Project Mercury,* SP-4201 (Washington: NASA, 1966), 115; Van Nimmen and Bruno, *NASA Resources 1958–1968,* 390.

15. James E. Webb, NASA Administrator, "Memorandum for the Vice President," May 23, 1961, NASA History Office, Washington.

16. Alfred Rosenthal, *Venture into Space: Early Years of Goddard Space Flight Center,* SP-4301 (Washington: NASA, 1968).

Chapter 2: Root Assumptions

Epigraph: House Committee on Science and Astronautics, Subcommittee on Manned Space Flight, *1965 NASA Authorization,* 88th Cong., 2d sess., 1964, 803.

1. Smith J. DeFrance, Paul F. Bikle, Harry J. Goett, Kurt H. Debus, Floyd L. Thompson, Abe Silverstein, Robert R. Gilruth, and Wernher von Braun; biography files, NASA History Office, Washington.

2. Schein, *Organizational Culture and Leadership,* 209–43.

3. Interview no. 36, August 12, 1987.

4. Interview no. 10, June 17, 1988.

5. Interview no. 9, September 11, 1987.

6. Interview no. 20, November 10, 1987.

7. Interview no. 47, February 19, 1988.

8. Roland, *Model Research,* vol. 2, 489; Hansen, *Engineer in Charge,* 413.

9. Interview no. 8, November 9, 1989.

10. Roland, *Model Research,* vol. 2, 423–26.

11. Interview no. 47, February 19, 1988.

12. NASA Office of Program Planning and Evaluation, "The Long Range Plan of the National Aeronautics and Space Administration," December 16, 1959, 5, NASA History Office, Washington.

13. George M. Low, Introduction to NASA, *What Made Apollo a Success?* SP-287 (Washington: NASA, 1971), 2–3. The chapters in this NASA publication also appeared in the March 1970 issue of *Aeronautics and Astronautics.*

14. Interview no. 9, September 11, 1987.

15. Interview no. 8, November 9, 1989.

16. Low, Introduction, 2–4.

17. See Roger F. Bilstein, *Stages to Saturn: A Technological History of the Apollo/ Saturn Launch Vehicle,* SP-4206 (Washington: NASA, 1980), 183–88.

18. Interview no. 46, February 23, 1988.

19. Interviewee letter to Sylvia D. Fries, June 25, 1990, 6, NASA History Office, Washington.

20. Interview no. 38, June 19, 1988.

21. Interview no. 15, March 21, 1988. The speaker notes that ABMA engineers put empty cans in the fuel and oxidizer tanks of the next Jupiter missile, already on the launch stand, as an emergency solution to prevent sloshing. Interviewee letter to the author, June 12, 1990.

22. Interview no. 42, May 10, 1989.

23. Interview no. 15, March 21, 1988.

24. Bilstein, *Stages to Saturn,* 123–25.

25. Interview no. 42, May 10, 1989.

26. Interview no. 10, June 17, 1988.

27. Interviewee letter to Sylvia D. Fries, 6.

28. Interview no. 8, November 9, 1989.

29. Interview no. 20, November 10, 1987.

30. Interview no. 38, June 19, 1988.

31. Wernher von Braun, "The Redstone, Jupiter, and Juno," in *The History of Rocket Technology*, edited by Eugene M. Emme (Detroit: Wayne State University Press, 1964). The rocket engine was manufactured by North American Aviation's Rocketdyne Division.

32. Ordway and Sharpe, *The Rocket Team*, chaps. 4, 5.

33. Bilstein, *Stages to Saturn*, 15, 81, 192–95, 265.

34. House Committee on Science and Astronautics, *Review of the Space Program*, 86th Cong., 2d sess., 1960, pt. 1, 400.

35. Interview no. 16, August 21, 1987.

36. Interview no. 15, March 21, 1988.

37. T. Keith Glennan, "The First Years of the National Aeronautics and Space Administration," vol. 1, Cleveland (1964), 7–8, 86, NASA History Office, Washington.

38. Arnold S. Levine, *Managing NASA in the Apollo Era*, SP-4102 (Washington: NASA, 1982), 70.

39. For an explanation of the contrasting systems, and a recommendation for more contractor latitude, see Advisory Committee on the Future of the U.S. Space Program, *Report of the Advisory Committee* (Washington: Government Printing Office, 1990), 40–41.

40. Interview no. 9, September 11, 1987.

41. House Committee on Science and Astronautics, Subcommittee on Manned Space Flight, *1965 NASA Authorization*, 803.

42. Bilstein, *Stages to Saturn*, 422; Courtney G. Brooks, James M. Grimwood, and Loyd S. Swenson, *Chariots for Apollo: A History of Manned Lunar Spacecraft*, SP-4205 (Washington: NASA, 1979), 409–11.

43. "Goddard Center—In House Plans," *Space Daily*, December 28, 1962; see also Rosenthal, *Venture into Space*, 47.

44. Interview no. 33, August 17, 1987.

45. McKinsey and Company, Inc., "An Evaluation of NASA's Contracting Policies, Organization, and Performance," contract NASW-144 (October 1960): pt. 2, 9, NASA Headquarters, Washington.

46. Interview no. 47, February 19, 1988.

47. Eberhard Rees, memo, December 9, 1965, attached to "Personal Impressions, Views and Recommendations," December 8, 1965; quoted in Bilstein, *Stages to Saturn*, 227.

48. Interview no. 53, August 19, 1987.

49. George V. Hanna, "Chronology of Work Stopages and Related Events, KSC/NASA and AFETR through July 1965," KSC Historical Report, October 1965, NASA History Office, Washington; Benson and Faherty, *Moonport*, 36–37.

50. Interview no. 12, March 22, 1988.

51. Interview no. 16, August 21, 1987.

52. National Academy of Public Administration, "NASA: Maintaining the Program Balance," vol. 1 (January 1991), 22–25.

53. Interview no. 46, February 23, 1988.

54. Interview no. 8, November 9, 1989.

55. Interview no. 36, August 12, 1987.

56. Interview no. 47, February 19, 1988.

57. Swenson, Grimwood, and Alexander, *This New Ocean*, 141–43.

58. Linda Neuman Ezell, *NASA Historical Data Book*, vol. 3, *Programs and Projects, 1969–1978*, SP-4012 (Washington: NASA, 1988), 152–64.

59. Interview no. 30, August 8, 1990.

60. Mark Craig interview, March 15, 1991.

61. Interview no. 47, February 19, 1988.

62. Interview no. 15, March 21, 1988.

63. Nimmen and Bruno, *NASA Historical Data Book*, 389–91.

64. For the creation of the engineering directorates and the evolution of project management as the space flight program moved from Langley to Houston, see the organization charts in Hansen, *Engineer in Charge*, 513–14; and Swenson, Grimwood, and Alexander, *This New Ocean*, 634–36.

65. See, for example, David Cleland and William R. King, *Systems Analysis and Project Management* (New York: McGraw-Hill, 1968), or Harvey M. Sapolsky, *The Polaris System Development* (Cambridge: Harvard University Press, 1972).

66. For an example of employee distribution at different centers, see National Academy of Public Administration, "NASA," 25–30.

67. House Committee on Science and Astronautics, Subcommittee on Manned Space Flight, *1965 NASA Authorization*, 1000.

68. Ibid., 803.

69. Eberhard Rees, "Project and Systems Management in the Apollo Program," in *Issues in NASA Program and Project Management*, edited by Francis T. Hoban, SP-6101(2) (Washington: NASA, 1989).

70. House Committee on Science and Astronautics, Subcommittee on NASA Oversight, *Apollo Program Management*, 91st Cong., 1st sess., 1969, 5.

71. Interview no. 20, November 10, 1987.

72. See J. D. Hodge, D. E. Fielder, and J. W. Roach, "Safety in Flight Operations," paper 67-824, prepared for the annual meeting of the American Institute of Aeronautics and Astronautics, October 23–27, 1967, NASA Headquarters Library, Washington, D.C.; and Charles Murray and Catherine Bly Cox, *Apollo: The Race to the Moon* (New York: Simon & Schuster, 1989), esp. 257–79.

73. See, for example, Christopher C. Kraft, John D. Hodge, and Eugene F. Kranz, "Mission Control for Manned Space Flight," paper prepared for the AIAA 2d Manned Space Flight Meeting, Dallas, Texas, April 23, 1963; and see Gene Kranz interview by James Burke, May 18, 1979; both in NASA History Office, Washington.

74. James E. Webb, memo to Mr. Shapley, "Subject: Problem of Adding 'Supervision' to the Activities of Top Management," September 19, 1967, NASA History Office, Washington.

75. See also National Academy of Public Administration, "NASA," 43–45.

76. Advisory Committee on the Future of the U.S. Space Program, *Report of the Advisory Committee*, 16.

77. Interview no. 5, August 31, 1990.

78. Interview no. 10, June 17, 1988.

79. National Aeronautics and Space Act of 1958 (72 Stat. 429); T. Keith Glennan, memo from the Administrator, "Subject: Establishment and Approval of Excepted Positions and Salaries," October 20, 1958, NASA History Office, Washington.

80. Levine, *Managing NASA in the Apollo Era*, 285 n. 13, and 110–13.

81. Advisory Committee on the Future of the U.S. Space Program, *Report of Advisory Committee*, 44–45; National Academy of Public Administration, "NASA," 43; National Commission on Public Service, *Leadership for America: Rebuilding the Public Interest* (Lexington, Mass.: Lexington, 1989), 251.

82. Social scientists commonly distinguish between "hygiene" factors that remove obstacles to motivation and the "motivators" that actually promote exceptional performance. See Frederick Herzberg, *Work and the Nature of Man* (New York: Thomas Y. Crowell, 1966).

83. Interview no. 38, June 19, 1988.

84. Interview no. 45, September 28, 1990.

85. Interview no. 6, August 20, 1987.

86. Interview no. 9, September 11, 1987.

87. Howard N. Braithwaite, paper to Willis Shapley, "History of Supergrade Positions, NACA/NASA," April 23, 1968, 1, 3, NASA History Office, Washington.

88. Interview no. 53, August 19, 1987.

89. House Committee on Science and Astronautics, "Statement of Dr. Wernher von Braun in Justification of Additional NASA-Excepted Positions," *Review of the Space Program*, 406.

90. Interview no. 12, March 22, 1988.

91. Interview no. 33, August 17, 1987.

92. Interview no. 37, June 23, 1988.

93. The National Academy of Public Administration reported in 1990 that NASA recruits fresh out of college had an aggregate grade point average of 3.2 but did not provide comparative data. National Academy of Public Administration, "NASA," 44.

94. Interview no. 5, August 31, 1990.

95. Interview no. 20, November 10, 1987. The average age of JSC flight controllers in 1991 was 32.7 years old (Johnson Space Center (Houston, Texas), Management Information System, Personal Statistics Application).

96. Interview no. 33, August 17, 1987, as amended by his June 25, 1990, letter to Sylvia Fries.

97. NASA, "Age Distribution of Permanent Employees (as of June 30, 1966)," NASA Headquarters, Personal Management Information System; NASA, "The Civil Service Work Force," issued annually since 1969 by the NASA Headquarters Office of Human Resources and Education, Washington.

98. Interview no. 20, November 10, 1987.

99. Interview no. 53, August 19, 1987.

100. Low, Introduction, 13.

101. George M. Low, "Remarks at Baccalaureate Ceremonies, Rensselaer Polytechnic Institute, June 1975," 3–4, NASA History Office, Washington. The *Apollo 13* accident is described in Henry S. F. Cooper, *13: The Flight That Failed* (New York: Dial, 1973).

102. Interview no. 36, August 12, 1987.

103. Interview no. 37, June 23, 1988.

104. Interview no. 6, August 20, 1987.

105. Interview no. 37, June 23, 1988.

106. Interview no. 20, November 10, 1987.

Chapter 3: Breaking Barriers

Epigraph: House Committee on Science and Astronautics, *1968 NASA Authorization*, 90th Cong., 1st sess., 1967, 6.

1. National Aeronautics and Space Act of 1958 (72 Stat. 427).

2. Interview no. 20, November 10, 1987.

3. Advisory Committee on the Future of the U.S. Space Program, *Report of the Advisory Committee*, 17.

4. Hallion, *On the Frontier*, 360.

5. Chuck Yeager and Leo Janos, *Yeager: An Autobiography* (New York: Bantam, 1985); Tom Wolfe, *The Right Stuff* (New York: Farrar, Straus, Giroux, 1979).

6. Interview no. 20, November 10, 1987.

7. Interview no. 9, September 11, 1987.

8. Interview no. 6, August 20, 1987.

9. Swenson, Grimwood, and Alexander, *This New Ocean*, 272–79. See also House Committee on Science and Astronautics, *1968 NASA Authorization*, 19–20.

10. Interview no. 6, August 20, 1987.

11. Interview no. 32, June 22, 1988.

12. Interview no. 38, June 19, 1988, The second flight of the Saturn V (unmanned) took place on April 4, 1968. With problems in all three stages of the giant rocket, NASA listed the test flight as unsuccessful. NASA, *Pocket Statistics*, January 1988, pt. B, 76; Bilstein, *Stages to Saturn*, 360–63; Murray and Cox, *Apollo*, 320–24.

13. Interview no. 38, June 19, 1988.

14. R. Cargill Hall, *Lunar Impact: A History of Project Ranger*, SP-4210 (Washington: NASA, 1977).

15. House Committee on Science and Astronautics, Subcommittee on NASA Oversight, *Investigation of Project Ranger*, 88th Cong., 2d sess., 1964.

16. Interview no. 16, August 21, 1987.

17. House Committee on Science and Astronautics, *1968 NASA Authorization*, 20. Also see Deputy Associate Administrator for Space Science and Applications to Associate Deputy Administrator, "Subject: Material for the Congressional Record," March 17, 1967, NASA History Office, Washington.

18. Interview no. 10, June 17, 1988.

19. Interview no. 33, August 17, 1987.

20. Sylvia D. Fries to William R. Graham, "Subject: Historical Summary: NASA: Safety Organization and Procedures in Manned Space Flight Programs," March 24, 1986, NASA History Office, Washington.

21. Interviewee letter to author, September 27, 1990.

22. Interview no. 36, August 12, 1987.

23. Interview no. 8, November 9, 1989.

24. Interview no. 37, June 23, 1988.

25. Interview no. 33, August 17, 1987.

26. Edward Clinton Ezell and Linda Neuman Ezell, *On Mars: Exploration of the Red Planet, 1958–1978*, SP-4212 (Washington: NASA, 1984).

27. Interview no. 33, August 17, 1987, as amended by his June 25, 1990, letter to Sylvia Fries.

28. Ibid.

29. See, for example, the work of the Apollo Crew Safety Review Board, in Brooks, Grimwood, and Swenson, *Chariots for Apollo*, 240–41, 265.

30. Interview no. 16, August 21, 1987.

31. Hodge, Fielder, and Roach, "Safety in Flight Operations"; Murray and Cox, *Apollo*, 261–63.

32. Barton C. Hacker and James M. Grimwood, *On the Shoulders of Titans: A History of Project Gemini*, SP-4203 (Washington: NASA, 1977), 256–62.

33. NASA, "NASA Major Launch Record," in NASA, *Pocket Statistics*, January 1988, (Washington: NASA Headquarters Office of Management, 1988), pt. B, 56–57.

34. Interview no. 6, August 20, 1987.

35. Interview no. 37, June 23, 1988.

36. Webb, memo to Mr. Shapley, "Subject: Problem of Adding 'Supervision' to the Activities of Top Management."

37. Interview no. 37, June 23, 1988.

38. Interview no. 9, September 11, 1987.

39. John W. Bullard, "History of the Redstone Missile," Historical Monograph Project Number AMC 23 M, U.S. Army Missile Command, Redstone Arsenal, Alabama, October 15, 1965.

40. Interview no. 12, March 22, 1988.

41. Interview no. 42, May 10, 1989.

42. Hans Frickner, telephone interview by John Low, October 11, 1991.

43. Interview no. 15, March 21, 1988.

44. See NASA, "The Civil Service Workforce," 1990, 84; also see similar tables in previous issues of this publication.

45. Interview no. 53, August 19, 1987.

46. Interview no. 16, August 21, 1987.

47. Interview no. 9, September 11, 1987.

48. National Aeronautics and Space Act of 1958 (72 Stat. 427).

49. Hugh L. Dryden, "Space Technology and the NACA" speech prepared for the Institute of Aeronautical Sciences, New York, January 27, 1958, NASA History Office, Washington; reprinted in *Aeronautical Engineering Review,* 17 (March 1958): 33.

50. Minutes, Research Steering Committee on Manned Space Flight, NASA Headquarters, May 25–26, 1959, 9–10, NASA History Office, Washington.

51. NASA Office of Program Planning and Evaluation, "The Long Range Plan of the National Aeronautics and Space Administration," table 1 and p. 28.

52. John M. Logsdon, *The Decision to Go to the Moon* (Cambridge, Mass.: MIT Press, 1970), 57.

53. Hugh L. Dryden, "The Exploration of Space," speech prepared for Cosmos Club, April 13, 1959, 7; NASA History Office, Washington.

54. Ibid.

55. Interview no. 20, November 10, 1987.

56. Interview no. 33, August 17, 1987, as amended by his June 25, 1990, letter to Sylvia Fries.

57. Interview no. 6, August 20, 1987.

58. NASA, *Spinoff* (Washington: Government Printing Office, 1990).

59. See NASA Lunar Landing Working Group (George M. Low, chairman), "A Plan for Manned Lunar Landing," February 7, 1961, 5; NASA History Office, Washington.

60. Interview no. 53, August 19, 1987.

61. Bilstein, *Stages to Saturn,* 14–15, 30, 91; Ordway and Sharpe, *The Rocket Team,* 370–71.

62. John L. Sloop, *Liquid Hydrogen as a Propulsion Fuel, 1945–1959,* SP-4404 (Washington: NASA, 1978); Bilstein, *Stages to Saturn,* 134–47.

63. Brooks, Grimwood, and Swenson, *Chariots for Apollo,* 154–55, 159, 300.

64. Interview no. 8, November 9, 1989.

65. Interview no. 33, August 17, 1987, as amended by his June 25, 1990, letter to Sylvia Fries.

66. Interview no. 38, June 19, 1988.

67. Interview no. 25, May 11, 1989.

68. Interview no. 14, June 15, 1988.

69. Interview no. 47, February 19, 1988.

70. Interview no. 27, June 21, 1988.

71. Interview no. 37, June 23, 1988.

72. *Webster's Third New International Dictionary of the English Language* (Springfield, Mass.: G. & C. Merriam, 1976). For a discussion of the behavior of professionals in government, also see Frederick C. Mosher, *Democracy and the Public Service* (New York: Oxford University Press, 1968).

73. Interviewee letter to Sylvia D. Fries, June 25, 1990.

74. Interview no. 4, June 20, 1988.

75. Interview no. 25, May 11, 1989.

76. Interview no. 33, August 17, 1987.

77. McDougall, . . . *The Heavens and the Earth.*

78. Interview no. 21, November 9, 1987.

79. W. R. Lucas, "Program Development Memorandum to Dr. Rees," June 16, 1970; NASA, "Space Shuttle Decisions," *NASA News,* Release 72-61, March 15, 1972; NASA, "Response to NBC Inquiry on Shuttle and Other Manned Space Flight Program Costs," March 10, 1981; all in NASA History Office, Washington. Also see John Logsdon, "The Decision to Develop the Space Shuttle," *Space Policy* 2 (May 1986): 103–19; and see Roger D. Launius, "The Development of the Space Shuttle, 1967–1972: Technological Innovation and Government Politics," paper prepared for the annual meeting of the Society for the History of Technology, Madison, Wisconsin, November 2, 1991.

80. Interview no. 47, February 19, 1988.

81. Interview no. 37, June 23, 1988.

82. Interview no. 32, June 22, 1988.

83. Executive Office of the President, Office of Science and Technology, President's Science Advisory Committee, "The Next Decade in Space," March 1970, 35–37, 51–52.

84. Interview no. 32, June 22, 1988.

85. Interview no. 33, August 17, 1987, as amended by his June 25, 1990, letter to Sylvia Fries.

86. Interview no. 24, March 22, 1988.

87. Interview no. 1, November 10, 1987.

Chapter 4: Becoming Conventional

Epigraph: Downs, *Inside Bureaucracy,* 18.

1. Interview no. 20, November 10, 1987.

2. Swenson, Grimwood, and Alexander, *This New Ocean,* 101.

3. Interview no. 20, November 10, 1987.

4. Hacker and Grimwood, *On the Shoulders of Titans,* 55–56, 105–16, 123–30.

5. Robert L. Rosholt, *An Administrative History of NASA, 1958–1963,* SP-4101 (Washington: NASA, 1966), 281–97.

6. Interview no. 45, September 28, 1990.

7. The 1961 reorganization replaced Abe Silverstein, a long-time NACA employee, with a project manager from the ballistic missile early warning system, D. Brainerd Holmes, as head of manned space flight activities at NASA headquarters. Brooks, Grimwood, and Swenson, *Chariots for Apollo,* 53–57, 127–30.

8. Interview no. 38, June 19, 1988.

9. Interview no. 32, June 22, 1988.

10. Interview no. 45, September 28, 1990.

11. Interview no. 6, August 20, 1987; letter from interviewee to author, October 15, 1990.

12. Neufeld, *The Development of Ballistic Missiles,* 110–13, 125–28, 140; Advisory Committee on the Future of the U.S. Space Program, *Report of the Advisory Committee,* 40–42.

13. Interview no. 45, September 28, 1990.

14. Joseph F. Shea, "Luncheon Presentation," speech prepared for the AIAA/NASA Conference on Innovative Technologies for the Exploration of Space, September 6, 1990, 4–5; NASA History Office, Washington.

15. Interview no. 12, March 22, 1988.

16. Interview no. 15, March 21, 1988.

17. Interview no. 32, June 22, 1988.

18. Interview no. 12, March 22, 1988.

19. House Committee on Science and Astronautics, "McDonnell Douglas Saturn/Apollo Program Management," *Apollo Program Management,* 66.

20. Interview no. 38, June 19, 1988.

21. See Bilstein, *Stages to Saturn,* 261–92.

22. Interview no. 38, June 19, 1988.

23. Interview no. 53, August 19, 1987.

24. See John R. Kimberly, Robert H. Miles, and associates, *The Organizational Life Cycle* (San Francisco: Jossey-Bass, 1987); Arthur L. Stinchcombe, "Social Structure and Organizations," in *Handbook of Organizations,* edited by James G. March (Chicago: Rand McNally, 1965); Downs, *Inside Bureaucracy;* and Bernstein, *Regulating Business by Independent Commission.*

25. McDougall, . . . *The Heavens and the Earth,* 312–15.

26. Herbert Kaufman, *Are Government Organizations Immortal?* (Washington: Brookings, 1976).

27. Selznick, *TVA and the Grass Roots;* David E. Lilienthal, *TVA: Democracy on the March* (New York: Harper & Brothers, 1953).

28. General Accounting Office, "Peace Corps: Meeting the Challenges of the 1990s,' NSIAD-90-122, May 1990. See also Kenneth J. Meier and John P. Plumlee, "Regulatory Administration and Organizational Rigidity," *Western Political Quarterly* 31 (March 1978): 86–88.

29. See Downs, *Inside Bureaucracy,* 14–15.

30. See Logsdon, *The Decision to Go to the Moon.*

31. Van Nimmen and Bruno, *NASA Historical Data Book,* 118; NASA, *Pocket Statistics,* January 1989, sec. C, 12.

32. "It has been estimated that it would cost the United States $40 billion—or an average of about $225 per person—to send a man to the moon. Would you like to see this amount spent for this purpose, or not?" Yes (33 percent); no (58 percent); no opinion (9 percent). George H. Gallup, *The Gallup Poll: Public Opinion 1935–1971,* vol. 3 (New York: Random House, 1972), 1720.

33. Herbert E. Krugman, "Public Attitudes toward the Apollo Space Program, 1965–1975," *Journal of Communication* 27 (Autumn 1977): 87–93; see also Michael A. G. Michaud, "The New Demographics of Space," *Aviation Space* 2 (Fall 1984): 46–47.

34. Elizabeth H. Hastings and Philip K. Hastings, eds., *Index to International Public Opinion, 1979–1980* (Westport, Conn.: Greenwood, 1981), 73.

35. Van Nimmen and Bruno, *NASA Historical Data Book,* 72–73.

36. Interview no. 53, August 19, 1987.

37. Interview no. 33, August 17, 1987.

38. National Aeronautics and Space Act of 1958 (72 Stat. 429–30); Levine, *Managing NASA in the Apollo Era,* 113–15.

39. Interview no. 33, August 17, 1987.

40. Interview no. 47, February 19, 1988; Rosholt, *An Administrative History of NASA, 1958–1963,* 79–81.

41. Interview no. 39, June 2, 1988. The story, while exaggerated, reflects the attitude of early space scientists toward government procedures. See John E. Naugle, *First Among Equals: The Selection of NASA Space Science Experiments,* SP-4215 (Washington: NASA, 1991), chap. 5.

42. Interview no. 39, June 2, 1988.

43. Interview no. 47, February 19, 1988.

44. Interview no. 30, August 8, 1990.

45. Advisory Committee on the Future of the U.S. Space Program, *Report of the Advisory Committee*, 42.

46. Levine, *Managing NASA in the Apollo Era*, 113–15.

47. Interview no. 30, August 8, 1990.

48. Herbert Kaufman, *Red Tape: Its Origins, Uses and Abuses* (Washington: Brookings, 1977).

49. Charles Perrow, *Complex Organizations* (Glenview, Ill.: Scott Foresman, 1972).

50. Max Weber, *From Max Weber: Essays in Sociology*, translated and edited by H. H. Gerth and C. Wright Mills (New York: Oxford University Press, 1946).

51. Robert K. Merton, "Bureaucratic Structure and Personality," *Social Forces* 18 (May 1940): 560–68.

52. President's Commission on the Accident at Three Mile Island, *Report of the President's Commission on the Accident at Three Mile Island* (New York: Pergamon, 1979), 52–53.

53. C. Northcote Parkinson, *Parkinson's Law and Other Studies in Administration* (Boston: Houghton-Mifflin, 1957).

54. See Charles T. Goodsell, *The Case for Bureaucracy*, 2d ed. (Chatham, N.J.: Chatham House, 1985).

55. See James Q. Wilson, *Bureaucracy* (New York: Basic 1989); and Gregory Lewis, "In Search of the Machiavellian Milquetoasts: Comparing Attitudes of Bureaucrats and Ordinary People," *Public Administration Review* 50 (March/April 1990): 220–27.

56. See John Child and Alfred Kieser, "Development of Organizations Over Time," in *Handbook of Organizational Design*, vol. 1, edited by Paul C. Nystrom and William H. Starbuck (New York: Oxford University Press, 1981), 44–51.

57. See, for example, NASA Office of Personnel Management, "The Civil Service Workforce," Fiscal Year 1989, NASA Headquarters, Washington, 80.

58. NASA Personnel Analysis and Evaluation Office, "The Civil Service Work Force," September 30, 1982, NASA Headquarters, Washington, 75.

59. Advisory Committee on the Future of the U.S. Space Program, *Report of the Advisory Committee*, 40–42.

60. Interview no. 32, June 22, 1988.

61. Steven M. Neuse, "TVA at Age Fifty—Reflections and Retrospect," *Public Administration Review* 43 (November/December 1983): 491–99. See also Edward F. Cox, Robert C. Fellmeth, and John E. Schulz, *"The Nader Report on the Federal Trade Commission* (New York: Richard W. Barton, 1969).

62. James Q. Wilson, ed., *The Politics of Regulation* (New York: Basic, 1980).

63. Downs, *Inside Bureaucracy*, 88.

64. Ibid., 20.

65. See Theodore J. Lowi, *The End of Liberalism* (New York: W. W. Norton, 1969); Grant McConnell, *Private Power and American Democracy* (New York: Knopf, 1966); and Selznick, *TVA and the Grass Roots*.

66. See Julius S. Brown, "Risk Propensity in Decision Making: A Comparison of Business and Public School Administrators," *Administrative Science Quarterly* 15 (1970): 473–81.

67. William Niskanen, *Bureaucracy and Representative Government* (Chicago: Aldine, 1971); and Dennis C. Mueller, *Public Choice II* (New York: Cambridge University Press, 1989), chap. 14. See also Niskanen, "Bureaucrats and Politicians," *Journal of Law and Economics* 18 (December 1975): 617–43.

68. NASA, *Spinoff*.

69. W. Warner Burke, "NASA Culture: Agency Report," August 1989 (New York: W. Warner Burke Associates), 21, 34, 37, 55.

70. Michel Crozier, *The Bureaucratic Phenomenon* (Chicago: University of Chicago Press, 1964).

71. Presidential Commission on the Space Shuttle Challenger Accident, *Report of the Presidential Commission*, 82–104; Advisory Committee on the Future of the U.S. Space Program, *Report of the Advisory Committee*, 16.

72. Interviewee letter to Sylvia Fries, June 25, 1990, 9.

73. Webb memo to Mr. Shapley, "Subject: Problem of Adding Supervision to the Activities of Top Management."

74. Interviewee letter to Sylvia Fries, June 25, 1990, 9.

75. Interview no. 36, August 12, 1987. The federal government established the National Railroad Passenger Corporation (Amtrack) in 1970 in order to rescue the failing railway passenger industry. The government corporation consumed federal subsidies and operated what many viewed as an inefficient railroad system.

76. Interview no. 6, August 20, 1987.

77. Kaufman, *Are Government Organizations Immortal?*

78. See Charles H. Levine, ed., *Managing Fiscal Stress* (Chatham, N.J.: Chatham Publishers, 1980).

79. James Q. Wilson, "Innovation in Organization: Notes Toward a Theory," in *The Management of Change and Conflict*, edited by John M. Thomas and Warren G. Bennis (New York: Penguin, 1972).

80. Kimberly, Miles, et al., *The Organizational Life Cycle*, 362–68.

81. James D. Thompson, *Organizations in Action* (New York: McGraw-Hill, 1967).

82. Aaron Wildavsky, *The New Politics of the Budgetary Process* (Glenview, Ill.: Scott, Foresman, 1988).

83. See Crozier, *The Bureaucratic Phenomenon*.

84. William Hines, "Moon Flight Plans Caused Houston's Surging Boom," *Washington Star*, November 18, 1962.

85. See "Institutional Assessment, Presentation to Dr. Fletcher, Dr. Low," July 2, 1975; "Headquarters Organization and Staffing Study Report," January 24, 1977; "Assessment of NASA Centers," September 13, 1977; all in Management Studies, 1970s notebooks, NASA History Office, Washington.

86. William R. Lucas, oral history interview by Andrew J. Dunar and Stephen P. Waring, June 19, 1989, quoted in the draft manuscript history of the Marshall Space Flight Center by Andrew J. Dunar et al., NASA History Office, Washington.

87. Dawson, *Engines and Innovation*; Elizabeth A. Muenger, *Searching the Horizon: A History of Ames Research Center, 1940–1976*, 148.

88. W. David Compton and Charles D. Benson, *Living and Working in Space: A History of Skylab*, SP-4208 (Washington: NASA, 1983), 377–78.

89. Smith, *The Space Telescope*.

90. Howard E. McCurdy, *The Space Station Decision: Incremental Politics and Technological Choice* (Baltimore: Johns Hopkins University Press, 1990), 75–80.

91. Dawson, *Engines and Innovation*, 205–7.

92. Smith, *The Space Telescope*, 228–30.

93. NASA, "Four NASA Centers Assigned Space Station Studies," *NASA News*, Release 84-85, June 29, 1984.

94. Philip Culbertson interview, December 11, 1987.

95. Thomas J. Lewin and V. K. Narayanan, "Keeping the Dream Alive: Managing the Space Station Program, 1982–1986," NASA Contractor Report 4272, July 1990; NASA History Office, Washington.

96. Advisory Committee on the Future of the U.S. Space Program, *Report of the Advisory Committee*, 20.

97. Ibid.

98. See Wildavsky, *The New Politics of the Budgetary Process*.

99. Synthesis Group on America's Space Exploration Initiative, *America at the Threshold* (Washington: Government Printing Office, 1991), 10–11; see also James Fisher and Andrew Lawler, "NASA, Space Council Split Over Moon-Mars Report," *Space News*, December 11, 1989; and Kathy Sawyer, "En Route to Space Goal, Groups Diverge," *Washington Post*, December 11, 1989.

100. David C. Morrison, "Out of Orbit," *National Journal* 24 (February 22, 1992): 456–57.

101. Kimberly, Miles, et al., *The Organizational Life Cycle*, 362–71.

102. Interview no. 38, June 19, 1988; Presidential Commission on the Space Shuttle Challenger Accident, *Report of the Presidential Commission*, 82–104, 199; S. C. Phillips, "Summary Report of the NASA Management Study Group: Recommendations to the Administrator," December 30, 1986, NASA History Office, NASA Headquarters, Washington; S. C. Phillips, "Effectiveness of NASA Headquarters: A Report for the National Aeronautics and Space Administration," Contract NASW-4254, National Academy of Public Administration, Washington, February 1988.

103. Interview no. 38, June 19, 1988; McCurdy, *The Space Station Decision*, 204–12.

104. Interview no. 38, June 19, 1988.

105. Interview no. 37, June 23, 1988.

106. Interview no. 32, June 22, 1988.

107. Presidential Commission on the Space Shuttle Challenger Accident, *Report of the Presidential Commission*, 104.

108. NASA Culture Survey, 1988, questionaire no. 431, author's files, Washington.

Chapter 5: Losing the Technical Culture

Epigraph: A. Hechtlinger, *Modern Science Dictionary* (New York: Gosset & Dunlap, 1959), 711.

1. Swenson, Grimwood, and Alexander, *This New Ocean*, 154–59.

2. See Glennan, "The First Years of the National Aeronautics and Space Administration," 7–8, 85–86.

3. NASA, "Annual Procurement Report," Fiscal Year 1978, NASA Headquarters, Washington, 78; Van Nimmen and Bruno, *NASA Historical Data Book*, 118.

4. Levine, *Managing NASA in the Apollo Era*, 86.

5. NASA, *Pocket Statistics*, 1989, sec. C., 12.

6. Procurement obligations increased as a percentage of total obligations from 83.4 percent in 1978 to 88.4 percent in 1989. NASA "Annual Procurement Report," issued annually.

7. Interview no. 20, November 10, 1987.

8. E. S. Groo, letter to William R. Lucas, December 1, 1975, 4, NASA History Office, Washington.

9. Interview no. 20, November 10, 1987.

10. Interview no. 32, June 22, 1988.

11. See Clarence H. Danhof, *Government Contracting and Technological Change* (Washington: Brookings, 1968).

12. Advisory Committee on the Future of the U.S. Space Program, *Report of the Advisory Committee*, 41.

13. Willis Hawkins, Robert Jastrow, William Nierenberg, and Frederick Seitz, "New Directions in Space: A Report on the Lunar and Mars Initiatives," George C. Marshall Institute, Washington, 1990, 24.

14. U.S. Commission on Organization of the Executive Branch of Government, *A Report to the Congress*, vol. 1, *Business Enterprises* (Washington: Government Printing Office, 1955), esp. 4–5 of the staff study on business enterprises outside of the Department of Defense.

15. Office of Management and Budget, "Performance of Commercial Activities," OMB Circular A-76, March 3, 1966, Washington.

16. See, for example, Niskanen, *Bureaucracy and Representative Government;* James M. Buchanan and Gordon Tullock, *The Calculus of Consent* (Ann Arbor: University of Michigan Press, 1962); Stuart Butler, *Privatizing Federal Spending* (Washington: Heritage Foundation, 1985); and E. S. Savas, *Privatizing the Public Sector* (Chatham, N.J.: Chatham House, 1982).

17. For a description of the traditional system, see House Committee on Science and Astronautics, Subcommittee on Manned Space Flight, *1970 NASA Authorization,* 91st Cong., 1st sess., 1969, 384–96.

18. Edward H. Kolcum, "Quality Gains New Emphasis after Shuttle Processing Review," *Aviation Week & Space Technology*, March 30, 1987.

19. Interview no. 20, November 10, 1987.

20. Interview no. 47, February 19, 1988.

21. Interview no. 32, June 22, 1988.

22. See Thompson, *Organizations in Action.*

23. Interview no. 37, June 23, 1988.

24. Congressional Budget Office, "The NASA Program in the 1990s and Beyond," May 1988.

25. See Advisory Committee on the Future of the U.S. Space Program, *Report of the Advisory Committee,* 38.

26. *Weekly Compilation of Presidential Documents,* Administration of Ronald Reagan, July 4, 1982, 870. NASA officials also proclaimed the space shuttle to be "a fully operational, reusable spacecraft." NASA, Mission Report, STS-4, "STS-4 Test Mission Simulates Operational Flight—President Terms Success 'Golden Spike' in Space," MS-004, no date, 1.

27. James M. Beggs, "Why the United States Needs a Space Station," speech prepared for the Detroit Economic Club and Detroit Engineering Society, June 23, 1982, 5, NASA History Office, Washington; reprinted under the same title in *Vital Speeches* 48 (August 1, 1982): 615–17.

28. Quoted from Rick Gore, "When the Space Shuttle Finally Flies," *National Geographic* 159 (March 1981): 317.

29. Interview no. 30, August 8, 1990.

30. Interview no. 32, June 22, 1988.

31. Interview no. 53, August 19, 1987.

32. Interview no. 16, August 21, 1987.

33. Interview no. 37, June 23, 1988.

34. White House officials downgraded their characterization of the space shuttle as "fully operational" before the *Challenger* accident. In 1984, the White House announced that the Space Transportation System would not be "fully operational" until at least 1988. "White House Fact Sheet: National Space Strategy," August 15, 1984, reproduced in NASA, *Aeronautics and Space Report of the President, 1984 Activities* (Washington: NASA, undated), 137.

35. S. C. Phillips, "Effectiveness of NASA Headquarters: A Report for the National Aeronautics and Space Administration," 27. See also, Advisory Committee on the Future of the U.S. Space Program, *Report of the Advisory Committee,* 38–40.

36. Interview no. 32, June 22, 1988.

37. Interview no. 8, November 9, 1989.

38. Interview no. 35, April 9, 1990.

39. Smith, *The Space Telescope*.

40. Interview no. 6, August 20, 1987.

41. Interview no. 16, August 21, 1987.

42. Interview no. 6, August 20, 1987.

43. See Charles Perrow, *Normal Accidents: Living with High-Risk Technologies* (New York: Basic, 1984).

44. Interview no. 8, November 9, 1989,

45. Interview no. 46, February 23, 1988.

46. Synthesis Group, *America at the Threshold*, 20–23.

47. Interview no. 6, August 20, 1987.

48. See Trudy E. Bell and Karl Esch, "The Space Shuttle: A Case of Subjective Engineering," *IEEE Spectrum* (June 1989): 44.

49. Interview no. 16, August 21, 1987.

50. R. P. Feynman, "Personal Observations on Reliability of Shuttle," in Presidential Commission on the Space Shuttle Challenger Accident, *Report of the Presidential Commission*, app. F, p. 1. Also see Bell and Esch, "The Space Shuttle."

51. National Academy of Public Administration, "NASA," 44.

52. Interview no. 8, November 9, 1989.

53. Interview no. 35, April 9, 1990.

54. A NAPA study discovered similar impressions. See National Academy of Public Administration, "NASA," 22–25.

55. Interview no. 16, August 21, 1987.

56. See Advisory Committee on the Future of the U.S. Space Program, *Report of the Advisory Committee*, 20, 44–46; Jay M. Shafritz, "An Indictment of NASA's Merit System," *Public Administration Review*, 52 (March/April 1992): 186–89.

Conclusion: Governmental Performance and Cultural Instability

Epigraph: John F. Kennedy, "Text of speech at Rice University, September 12, 1962," in *"Let the Word Go Forth": The Speeches, Statements, and Writings of John F. Kennedy, 1947–1963*, edited by Theodore C. Sorensen (New York: Delacorte, 1988), 178.

1. National Aeronautics and Space Act of 1958 (72 Stat. 427).

2. John F. Kennedy, "Special Address to Congress on Urgent National Needs, May 25, 1961," in Sorensen, *"Let the Word Go Forth,"* 174.

3. For an analysis of the role of failure in successful design, see Henry Petroski, *To Engineer Is Human* (New York: St. Martin's, 1985).

4. Rosabeth Moss Kanter, *The Change Masters* (New York: Simon & Schuster, 1983).

5. LaPorte and Consolini, "Working in Practice but Not in Theory," 32. See also John Pfeiffer, "The Secret of Life at the Limits: Cogs Become Big Wheels," *Smithsonian* 20 (July 1989): 38–48.

6. Statement of George M. Low before the House Committee on Science and Technology, Subcommittee on Energy Research and Production, May 24, 1979, NASA History Office, Washington.

7. See, for example, Richard Rhodes, *The Making of the Atomic Bomb* (New York: Simon & Schuster, 1986).

8. Peters and Waterman, *In Search of Excellence*.

9. Orion White, "The Dialectical Organization: An Alternative to Bureaucracy," *Public Administration Review* 29 (January/February 1969): 32–42.

10. Ott, *The Organizational Culture Perspective*, 4–5, 55–56, 95–96.

11. Schein, *Organizational Culture and Leadership*, 267.

12. Office of Management and Budget, *Historical Tables: Budget of the United States Government, Fiscal Year 1989* (Washington: Government Printing Office, 1988), 89.

13. House Committee on Science and Astronautics, Subcommittee on Manned Space Flight, *1974 NASA Authorization*, 93d Cong., 1st sess., 1973, 1270–74.

14. Ezell, *NASA Historical Data Book*, vol. 2, 205; vol. 3, 135.

15. Ibid., vol. 2, 205.

16. Smith, *The Space Telescope*, 371.

17. Advisory Committee on the Future of the U.S. Space Program, *Report of the Advisory Committee*, 40.

18. Glennan, "The First Years of the National Aeronautics and Space Administration," 86.

19. Levine and Narayanan, "Keeping the Dream Alive," 32.

20. Advisory Committee on the Future of the U.S. Space Program, *Report of the Advisory Committee*, 41.

21. See Bernard Rosen, *Holding Government Bureaucracies Accountable* (New York: Praeger, 1989).

22. See Leon Festinger, *A Theory of Cognitive Dissonance* (Stanford: Stanford University Press, 1957); and Irving Janis, *Victims of Groupthink* (Boston: Houghton Mifflin, 1972).

23. See Hall, *Lunar Impact: A History of Project Ranger*, 246–255.

24. See Frederick C. Mosher, *Democracy and the Public Service* (New York: Oxford University Press, 1968); and Weber, "Essay on Bureaucracy," *From Max Weber*.

25. Ethics in Government Act of 1978 (92 Stat. 1824); Crimes and Criminal Procedure: Disqualification of Former Officers and Employees (18 U.S. Code 207).

26. Downs, *Inside Bureaucracy*, chap. 2.

27. See Henry Mintzberg, *Power in and around Organizations* (Englewood Cliffs, N.J.: Prentice-Hall, 1983); and Jeffrey Pfeffer, *Power in Organizations* (Marshfield, Mass.: Pitman, 1981).

28. See S. C. Phillips, "Summary Report of the NASA Management Study Group"; and Lewin and Narayanan, "Keeping the Dream Alive."

29. Brooks, Grimwood, and Swenson, *Chariots for Apollo*, 71.

30. See, for example, Perrow, *Complex Organizations*, 1–57.

31. See Wilson, *Bureaucracy*, chap. 9.

32. Sapolsky, *The Polaris System Development*.

33. Synthesis Group, *America at the Threshold*, 8.

34. Interview no. 20, November 10, 1987.

Index

169–70; "excepted" positions, 52, 107; expectations of, 50–51, 56–58; evidence for, 56; quality of people who formed NASA, 51–52; relationship to program success, 60; technical quality, 157

Expansion-contraction cycle, 99–100, 125; in NASA, 101–6, 136–38

Explorer satellite program, 21, 44; *Explorer 1*, 16–17

Ezell, Linda Neuman, 191

F-1 rocket engine, 33, 63

Faget, Max, 46–48

Faherty, William B., 186

Failure, 3, 58, 63–65, 69, 95–96, 123, 161, 165, 186; interactive, 153–54; methods for reducing, 65–69, 151, 153–54; failure rates, 69, 150; successful versus complete failure, 32; tolerance of, 70, 149–52, 154–55, 164, 166, 168. *See also* Risk taking

Federal Maritime Commission, 118

Federal Trade Commission, 118

Feynman, Richard, 154

Flexibility, 38, 57, 106–11, 121–24, 126–28, 163

Flight tests. *See* Testing

Flying in pairs, 68, 151

Food and Drug Administration, 118

Forest Service, 5, 188

Fries, Sylvia D. (now Sylvia K. Kraemer), 187

Frontier mentality, 71–77, 145, 161

Frost, Larry F., 187

Fuel cell, 75

Galileo probe, 3

Gemini program, 63, 68–69, 147, 150–51, 186; management problems, 90

Generations, 78–85, 161

German rocket team, 32–33, 46, 54–55, 70–71, 95, 186–87; culture of, 16–17, 93–94. *See also* Army Ballistic Missile Agency; Marshall Space Flight Center

Gilmour, Robert, 188

Gilruth, Robert, 25, 46, 60

Glennan, T. Keith, 25, 38, 56, 167

Global Low Orbiting Message Relay Satellite (GLOMR), 148

Goal displacement, 113. *See also* Bureaucracy: bureaucratic dysfunctions

Goddard Space Flight Center, 51, 108, 127, 143; culture of, 20–21; hands-on work, 44–45; in-house norms, 39; site selection, 107–8

Government organizations: tendencies toward fragmentation, 5

Grade creep, 117

Greenbelt, Maryland, 20, 108

Grimwood, James M., 186

Grissom, Virgil, 69

Groves, Leslie R., 163

Hacker, Barton C., 186

Hallion, Richard P., 186

Hands-on work, 34, 42–49, 156, 160, 165; advantages of, 46, 48; in Explorer satellite program, 44; loss of, 156–57; in Project Mercury, 44; relationship between technical laboratories and project managers, 46–48; relationship to in-house capability, 42–43

Hansen, James R., 186

Hard work, 50, 56–57

Headquarters staff, 27; growth of, 114–15

High-performance organizations, 50–51, 158, 161, 163, 166; instability of, 1–2, 159, 165, 172–73

High-reliability organizations, 6, 94, 161–63, 188

High Speed Flight Research Station, 13, 186

Holmes, Brainerd, 88

Hubble Space Telescope, 3, 21, 23, 38, 127, 143, 152, 167, 187; testing, 149–50

Hughes Communications, 148, 151

Hydrogen (liquid) fueled rockets, 14

IBM, 38

Incrementalism, 77–78, 94

Industry. *See* Contracting

In-house technical capability, 34–42, 44–45, 92–93, 135, 141, 160, 165, 167–68, 187; and attitude toward contractors, 40–41; loss of, 136–39, 156–57; machine and fabrication shops, 36, 43, 45; mix of employees, 136–38; reasons for, 37–38, 40; relationship to hands-on work, 42–43; satellite projects, 39; 10 percent rule, 39, 140; as training for project managers, 35–36, 43, 93. *See also* Contracting

Innovation. *See* Change

hands-on tradition, 43–44; in-house
tradition, 35–36, 134; technical papers,
26–27; test philosophy, 26–28
National Aeronautics and Space Administration (NASA), 61; as advocate for
space flight, 73–74; age of work force,
104–5; changes summarized, 1–3, 91,
163–65, 172; executive leadership,
87–89, 120; factors contributing to program success, 3–4; long-range plan,
29, 74; oral traditions, 41, 70–71; promotion opportunities, 105–6; reductions in force, 104; reputation of, 2–3;
subcultures, 19–23, 92, 171–72
—organization: Human Resources and Education Office, 191; NASA History Office, 186, 191; Office of Program Planning and Evaluation, 74. See also names
of specific field centers
National Radiation Laboratory, 173
National Security Space Program (Department of Defense), 168
National Space Council, 110, 129
Naval aircraft carrier study, 162
Naval Research Laboratory, 11, 15, 17,
20–21, 51
Navy satellite program, 15–16
Niskanen, William, 119
North American Aviation, 41, 66
Nuclear Regulatory Commission, 113

O'Connor, Edmund F., 92
Office of Management and Budget, 139
Open communication, 65–68, 70–71,
98, 121–23, 131, 154–55, 157–58,
163
Operations. See Space flight operations
Oppenheimer, J. Robert, 163
Orbiting Astronomical Observatories, 21,
149–50, 166
Orbiting Geophysical Observatories, 21,
65
Ordway, Frederick I., 187
Organizational culture, 185, 187–88,
191–92; cultural politics, 24, 171–72;
defined, 4; functions of, 4–5; insulation from outside pressures, 168; methods for studying, 8; persistance of, 24,
159, 164, 173–74; relationship to performance (organizational excellence),
5–7, 163, 187–88; role of charismatic
leaders, 25; subcultures, 7–8
Ott, J. Stephen, 7, 187
Ouchi, William, 5–6, 187

Paperwork, 40, 42, 72, 117; paper trails,
59–60. See also Administrative regula-
tions
Parkinson, C. Northcote, 113
Parkinson's Law, 113
Pay (salaries in government), 26, 52–54,
107, 110
Peace Corps, 100
Peenemünde, 16, 36
Peters, Thomas, 6, 163, 187
Petrone, Rocco, 19
Phillips, Samuel C., 60, 92, 97
Pioneer spacecraft, 23
Plumlee, John P., 188
Planning for success, 123
Pogo effect, 63
Polaris missile development program, 173
Politics and bureaucracy, 88, 157; con-
gressional oversight, 110–11; NASA's
political environment, 165–68; political
considerations in decision making, 82,
85–87; political oversight, 109,
123–24. See also Technical criteria for
decision making
Presidential Commission on the Space
Shuttle Challenger Accident (Rogers
Commission), 19, 130, 187
President's Science Advisory Committee
(PSAC), 86–87
Privatization, 139
Professional administrators, 114–17
Professionalism in government, 81–82,
112, 173–74
Program management, 186; centralized
approach, 91–92, 96–97; "lead center"
approach, 129–31, 163
Project management, 57; relationship to
technical laboratories, 46–48
Project selection, 108–9, 125
Promotions, 116–17
Public opinion, 102–3

Radar, 173
Ranger project, 3, 21, 64, 123, 155,
167–68, 186
Reagan, Ronald, 144
Recruitment of employees who "fit the
mold," 78, 87, 165, 173
Redstone Arsenal, 16, 36
Redstone rocket, 32, 36, 42, 70–71, 77
Reliability. See High-reliability organiza-
tions
Research and development: versus opera-
tions, 72–73, 142–43, 144–46

Designed by Sally Harris / Summer Hill Books
Composed by Achorn Graphics
in Meridien with Eras Bold and Demi Italic
Printed by BookCrafters
on 50-lb. Booktext Natural